Martin Sternstein, Ph.D.
Professor of Mathematics
Ithaca College
Ithaca, New York

Statistics

All inquiries should be addressed to:
Barron's Educational Series, Inc.
250 Wireless Boulevard
Hauppauge, New York 11788

Library of Congress Catalog Card No. 94-4069

International Standard Book No. 0-8120-1869-9

Library of Congress Cataloging-in-Publication Data
Sternstein, Martin.
  Statistics / Martin Sternstein.
    p.   cm. — (Barron's EZ 101 study keys)
  Includes index.
  ISBN 0-8120-1869-9
  1. Statistics—Study and teaching.   I. Title. II. Series.
QA276.18.S74   1994
519.5—dc20   94-4069
                                                      CIP

PRINTED IN THE UNITED STATES OF AMERICA
19 18 17 16 15 14 13 12 11

CONTENTS

To the reader

This book is intended as an overview and supplementary learning aid for college students taking an introductory course in statistics or needing a self review guide of elementary statistics as a prerequisite for another course. You should be able to *self-read* this guide and find it useful when doing homework or studying for exams, and in clearing up basic points when the standard textbook might use extensive theoretical discussion.

The material is divided into nine major **Themes** and then further into sixty-four units called **Keys** each focusing on a single topic. Objectives are made very explicit, and there are carefully worked out examples. Special emphasis has been given to word problems showing the usefulness of statistics in a variety of disciplines.

Some of the problems especially those involving binomial and Poisson probabilities, require a hand calculator with an x^y-key. A few of the problems involve small data sets, and a "stat" mode on your calculator will shorten calculations involving population and sample standard deviations, σ and s, or σ_n and σ_{n-1} respectively. And of course if you are working on a large data project, access to a computer statistics package is almost a necessity. This review book will help you with the basic concepts, language, and techniques of statistics so as to better understand the computer calculations.

I wish to thank my colleagues who suggested the topics; my students who tested and commented on the examples; my sons Jonathan and Jeremy who gave up "game time" on the family computer so that their dad could work on this book; and my wife Faith without whose love and encouragement this project would not have been finished on time.

Ithaca College
August 1994

Martin Sternstein

Theme 1 DESCRIPTIVE STATISTICS

*T*he need to make sense of masses of information has led to formalized ways of describing the tremendous and ever-growing quantity of numerical data being collected in almost all areas of knowledge. Given a raw set of data, there is often no apparent overall pattern. Perhaps some values are more frequent, sometimes a few extreme values stand out, and usually the range of values is noticeable. Presenting data involves such concepts as representative or average values, measures of dispersion, and positions of various values, all of which fall under the broad topic of *descriptive statistics*, the subject of Theme 1. This is in contrast to *statistical analysis*, which draws inferences from limited data, a subject that will be discussed in later themes.

INDIVIDUAL KEYS IN THIS THEME

1	Central tendency
2	Variability
3	Position
4	The empirical rule
5	Chebyshev's theorem
6	Theme exercises with answers

Key 1 Central tendency

OVERVIEW   *The word* average *arises in phrases common to everyday conversation, from* batting averages *to* average life expectancies, *and has come to mean a "representative score" or a "typical value" or the "center" of a distribution. Mathematically, there are a variety of ways to define the average of a set of data, and in this key we consider the three most common methods.*

The median: Derived from the Latin *medius,* meaning "middle," the **median** is the middle number of a set of numbers arranged in numerical order. (If there is an even number of values, the median is the result of adding the two middle values and dividing by 2.)

- The median of the set {2, 3, 6, 6, 7, 9, 10, 13, 25} is 7.
- The median of the set {2, 3, 4, 7, 8.4, 9, 35, 46} is $\frac{1}{2}$ (7 + 8.4) = 7.7.

  The median is not affected by exactly how large the larger values are or by exactly how small the smaller values are. Thus it is a particularly useful measurement when the extreme values, called *outliers*, are in some way suspicious, or if one wants to diminish their effect. For example, if ten mice try to solve a maze, and nine take under 15 minutes, while one is still trying after 24 hours, then the most representative value is the median. In certain other situations the median offers the most economical and quickest technique to calculate an average. For example, suppose 10,000 light bulbs of a particular brand are installed in a factory. An average life expectancy for the bulbs can most easily be obtained by noting how much time passes before exactly one-half of them have had to be replaced. The median is also useful in certain kinds of medical research. For example, to compare the relative strengths of different poisons, a scientist notes what dosage of each poison will result in the deaths of exactly one-half the test animals. This median lethal dose is not influenced by one of the animals being especially susceptible to a particular poison.

The mode: The **mode**, or most frequent value, is an easily understood representative score. It is clear what is meant by "the most common family size is 4" or "the professor gives out more B's than any other grade."

- The mode of {2, 3, 6, 6, 7, 9, 10, 13, 25} is 6.
  When two scores have equal frequency, and this frequency is higher than any other, we say that the set is *bimodal*.
- The set {2, 3, 6, 6, 7, 7, 10, 13, 25} is bimodal, with modes 6 and 7.

The mean: Summing the values or scores, and dividing by the number of values or scores, gives the **mean**, the most important measure of central tendency for statistical *analysis*.

- {2, 3, 6, 6, 7, 9, 10, 13, 25} has mean $(2 + 3 + 6 + 6 + 7 + 9 + 10 + 13 + 25)/9 = 81/9 = 9$.
  The mean of a *whole population* (complete set of items of interest) is often denoted by the Greek letter μ (mu), while the mean of a *sample* (a part of a population) is often denoted as $\bar{x}$.
- The mean average value of the set of all houses in the United States might be μ = \$56,400, while the mean average of 100 randomly chosen houses might be $\bar{x}$ = \$52,100 or perhaps $\bar{x}$ = \$63,800, or even $\bar{x}$ = \$124,000.

In statistics one learns how to estimate a population mean from a sample mean. Throughout this book, sample implies *random* sample; that is, the sample must be selected under conditions such that each element of the population has an equal chance to be included in the sample. In the real world, this process of *random* selection is often very difficult to achieve.

Notation: μ = $\bar{x}$ = $(\sum x)/n$, where $\sum x$ represents the sum of all the elements of the set under consideration, while n is the actual number of elements ($\sum$ is the upper-case Greek letter sigma).

Properties: Unlike the median or mode, the mean is sensitive to a change in any value. Adding the same constant to each value will increase the mean by a like amount. Multiplying each value by the same constant will multiply the mean by a like amount. If a sum is formed by selecting one element from each of two sets, then the mean of all such sums is simply the sum of the means of the two sets.

Trimmed mean: This variation of the mean is an attempt to reduce the influence of extreme values (outliers) on the mean. It is calculated by arranging the terms in numerical order, taking away the first quarter and the fourth quarter of the values, and then finding the arithmetic mean of what remains.

Key 2   Variability

OVERVIEW   *In describing a set of numbers, not only is it useful to designate an average score, but also it is important to be able to indicate the **variability** or the **dispersion** of the measurements. A producer of time bombs aims for small variability—it would not be good if his 30-minute fuses actually ranged from 10 minutes to 50 minutes before detonation. On the other hand, a teacher interested in distinguishing the better from the poorer students aims to design exams with large variability in results—it would not be helpful if all the students scored exactly the same. The players on two basketball teams might have the same average height, but this fact doesn't tell the whole story—if the dispersions are quite different, one team might have a 7-foot player, whereas the other might have no one over 6 feet. Two Mediterranean holiday cruises might advertise the same average age for their passengers, but one could have only passengers between 20 and 25 years old, while the other had only middle-aged parents in their 40's together with their children under age 10.*

The range: The difference between the largest and smallest values of a set is called the **range**. While it is the most easily calculated measure of variability, the range is entirely dependent on two extreme values and is insensitive to what is happening between. One use of the range is to consider samples with very few items. For example, some quality-control techniques involve taking periodic small samples and basing further action upon the range found in several samples.

The variance: This measure of variability indicates dispersion around the mean. The **variance**, denoted by σ^2 (σ is the lower-case Greek letter sigma), is the average of the squared deviations from the mean:

$$\sigma^2 = \frac{\Sigma(x-\mu)^2}{n}$$

For circumstances specified in Key 29, the variance of a sample, denoted as s^2, is calculated as follows:

$$s^2 = \frac{\Sigma(x-\bar{x})^2}{n-1}$$

The standard deviation: Typically represented by σ or s, the **standard deviation** is the *square root* of the variance.

- If $X = \{2, 9, 11, 22\}$, then

$$\mu = \frac{2 + 9 + 11 + 22}{4} = 11$$

$$\sigma^2 = \frac{(2-11)^2 + (9-11)^2 + (11-11)^2 + (22-11)^2}{4} = 51.5$$

and

$$\sigma = \sqrt{51.5} = 7.176$$

An alternative arithmetical tool for calculating the variance and standard deviation comes from these equations:

$$\sigma^2 = \frac{\Sigma x^2}{n} - \mu^2 \quad \text{and} \quad s^2 = \frac{\Sigma x^2 - \dfrac{(\Sigma x)^2}{n}}{n-1}$$

- For the above $\{2, 9, 11, 22\}$ we could also have calculated

$$\Sigma x^2 = 2^2 + 9^2 + 11^2 + 22^2 = 690$$

so

$$\sigma^2 = \frac{690}{4} - 11^2 = 172.5 - 121 = 51.5$$

The interquartile range: This is one method of removing the influence of extreme values on the range. It is calculated by arranging the data in numerical order, removing the upper and lower one-quarter of the values, and then noting the range of the remaining values.

Relative variability: A comparison of two variances may be more meaningful if the means of the populations are also taken into consideration. **Relative variability** is defined to be the quotient obtained by dividing the standard deviation by the mean. Usually it is then expressed as a percentage.

Key 3 Position

OVERVIEW *We have seen several ways of choosing a value to represent the center of a distribution. It is also important to be able to talk about the position of any other value. In some situations, such as wine tasting, one is interested in simple rankings. Other cases, for example, evaluating college applications, may involve positioning according to percentile rankings. There are also situations in which we are able to specify position by making use of measurements of both central tendency and variability, that is, by the z-score.*

Simple ranking: Simple ranking, which involves arranging the elements in some order and noting where in the order a particular value falls, is straightforward. We know what it means for someone to graduate second in a class of 435, or for a player from a team of size 30 to have the seventh best batting average. Simple ranking is useful even when no numerical values are associated with each element. For example, detergents may be ranked according to relative cleansing ability without any numerical measurements of strength.

Percentile ranking: Percentile ranking, which indicates what percent of all scores fall below the value under consideration, is helpful for comparing positions with different bases. We can easily compare a rank of 176 out of 704 with a rank of 187 out of 935 by noting that the first equals a percentile rank of 75%, the second a rank of 80%. Percentile rank is also useful when the exact population size is not known or is irrelevant. For example, it is more meaningful to say that a student scored in the 90th percentile on a national exam, rather than trying to determine an exact ranking among some large number of test takers.

The z-score: The z-score is a measure of position that takes into account both the center and the dispersion of the distribution. More specifically, the z-score of a value tells how many standard deviations the value is from the mean. Mathematically, $x - \mu$ gives the raw distance from μ to x; dividing by σ converts this distance to numbers of standard deviations. Thus $z = (x - \mu)/\sigma$, where x is a raw score, μ is the mean, and σ is the standard deviation. Note that, if the score x is greater than the mean μ, then z is positive, if less, then z is negative.

Given a z-score, we can also reverse the procedure and find the corresponding raw score. Solving for x gives: $x = \mu + z\sigma$.

KEY EXAMPLE

Suppose that the average (mean) price of gasoline in a large city is $1.80 per gallon with a standard deviation of $0.05. Then $1.90 has a z-score of $(1.90 - 1.80)/0.05 = +2$, while $1.65 has a z-score of $(1.65 - 1.80)/0.05 = -3$. Alternatively, a z-score of $+2.2$ corresponds to a raw score of $1.80 + 2.2(0.05) = 1.80 + 0.11 = 1.91$, while a z-score of -1.6 corresponds to $1.80 - 1.6(0.05) = 1.72$. It is often useful to portray integer z-scores and the corresponding raw scores as follows:

1.65	1.70	1.75	1.80	1.85	1.90	1.95	price/gallon
−3	−2	−1	0	1	2	3	z-score

KEY EXAMPLE

An assembly line produces an average of 12,600 units per month with a standard deviation of 830 units. Adding and subtracting multiples of 830 to the mean 12,600 gives:

10110	10940	11770	12600	13430	14260	15090	units/month
−3	−2	−1	0	1	2	3	z-score

11,106 has a z-score of $(11{,}106 - 12{,}600)/830 = -1.8$, while a z-score of 2.4 corresponds to $12{,}600 + 2.4(830) = 14{,}592$

Key 4 The empirical rule

OVERVIEW *The **empirical rule** applies specifically to symmetric, "bell-shaped" data. In this case, about 68% of the values lie within one standard deviation of the mean, about 95% of the values lie within two standard deviations of the mean, and more than 99% of the values lie within three standard deviations of the mean.*

In terms of z-scores we have the following displays:

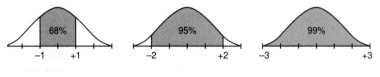

KEY EXAMPLE

Suppose that taxi cabs in New York City are driven an average of 75,000 miles per year with a standard deviation of 12,000 miles. Assuming that the distribution is "bell-shaped," we can conclude that approximately 68% of the taxis are driven between 63,000 and 87,000 miles per year, approximately 95% are driven between 51,000 and 99,000 miles, and virtually all are driven between 39,000 and 111,000 miles.

The empirical rule also gives a useful quick estimate for the standard deviation in terms of the range. We can see in the display above that 95% of the data fall within a span of four standard deviations (from –2 to +2 on the z-score line), and 99% of the data fall within a span of six standard deviations (from –3 to +3 on the z-score line). It is therefore reasonable to conclude that the standard deviation is roughly between one-fourth and one-sixth of the range. Since we can find the range of a set almost immediately, this technique for estimating the standard deviation is often helpful also in pointing out probable arithmetic errors.

KEY EXAMPLE

If the range of a data set is 60, the standard deviation should be expected to be between $(1/6)60 = 10$ and $(1/4)60 = 15$. If the standard deviation is calculated to be 0.32 or 87, there is a probable arithmetic error; a calculation of 12, however, is reasonable.

Key 5 Chebyshev's theorem

OVERVIEW *When data are spread out, the standard deviation is larger; when data are tightly compacted, the standard deviation is smaller. However, no matter what the dispersion, and even if the data are not "bell-shaped," certain percentages of the data will always fall within specified numbers of standard deviations from the mean.*

The Russian mathematician Chebyshev showed that, for any set of data, *at least* $1 - 1/k^2$ of the values lie within k standard deviations of the mean. As a percent, at least $100(1 - 1/k^2)\%$ of the values are within k standard deviations of the mean. In terms of z-scores, at least $1 - 1/k^2$ of the values have z-scores between $-k$ and $+k$. Therefore, when $k = 3$, at least $1 - 1/9$ or 88.89% of the values lie within three standard deviations of the mean. And, when $k = 5$, at least $1 - 1/25$ or 96% of the values have z-scores between -5 and $+5$.

KEY EXAMPLE

Suppose an electronic part takes an average of 3.4 hours to move through an assembly line with a standard deviation of 0.5 hour. Then, using $k = 2$, we find that at least $1 - 1/4$ or 75% of the parts take between $3.4 - 2(0.5) = 2.4$ hours and $3.4 + 2(0.5) = 4.4$ hours to move through the line. Similarly, with $k = 4$ at least 15/16 or 93.75% of the parts take between 1.4 and 5.4 hours.

Given a range around the mean, we can convert to z-scores and ask about the percentage of values in the range. If one set has a mean $\mu = 85$ and standard deviation $\sigma = 1$, while a second set has $\mu = 85$ and $\sigma = 5$, then, for example, the percentages of values between 75 and 95 will be different. For the first set, the relevant z-scores are ± 10, while for the second the relevant z-scores are ± 2. Thus 99/100 or 99% of the first set's values, but only 3/4 or 75% of the second set's values, lie between 75 and 95.

KEY EXAMPLE

Suppose the daily intake at a toll booth averages \$3500 with a standard deviation of \$200. What percentage of the daily intakes should be between \$3000 and \$4000?

Answer: The relevant z-scores are $\pm 500/200 = \pm 2.5$, so Chebyshev's theorem says that at least $1 - 1/(2.5)^2 = 21/25$ or 84% of the daily intakes should be between $3000 and $4000.

Note that, for $k = 1$, $1 - 1/k^2 = 0$; here the theorem gives no useful information.

Reminder: The power of Chebyshev's theorem is that it applies to **all** sets of data. However, if the data are "bell-shaped," we can draw stronger conclusions by using the empirical rule, stated in Key 4.

KEY EXAMPLE

Shape will be discussed in Theme 2, but intuitively consider the following two sets of data:

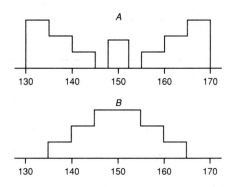

True or false?
    Both sets have about the same mean.
    The variance of set A is greater than the variance of set B.
    Chebyshev's theorem applies to both sets.
    The empirical rule applies only to set A.
    You can be sure that the standard deviation of set A is greater than 5.

Answers: T (both are about 150), T, T (Chebyshev's theorem applies to all sets), F (the empirical rule applies only to bell-shaped data like set B), T (by Chebyshev's theorem at least 75% of the data is within two standard deviations of the mean, but only a small portion of set A is between 140 and 160).

Key 6   Theme exercises with answers

OVERVIEW   *Sample questions of the type that might appear on homework assignments and tests are presented with answers*

- Find the median, mode, and mean of these two sets: {2, 8, 33, 2, 20} and {2, 8, 58, 8, 20}. What principle is illustrated?
  Answer: Remember to first arrange the data in numerical order to find the median.

    For {2, 2, 8, 20, 33} the median is 8, the mode is 2, and the mean is $(2 + 2 + 8 + 20 + 33)/5 = 13$.

    For {2, 2, 8, 20, 58} the median is 8, the mode is 2, and the mean is $(2 + 2 + 8 + 20 + 58)/5 = 18$.

  This example illustrates that the mean, unlike the median and mode, is sensitive to any change in value.

- Calculate the mean of each of these sets: {3, 7, 15, 22, 23, 38}, {6, 10, 18, 25, 26, 41}, and {6, 14, 30, 44, 46, 76}. What principle is illustrated?
  Answer: The three resulting means are 18, $18 + 3 = 21$, and $18 \times 2 = 36$. Adding or multiplying the same constant to each value will do the same to the mean.

- What are the range and the variance of {2, 4, 4, 5, 7, 8}?
  Answer: The range = $8 - 2 = 6$.

    The mean $\mu = (2 + 4 + 4 + 5 + 7 + 8)/6 = 5$.

    The variance $\sigma^2 = [(2 - 5)^2 + (4 - 5)^2 + (4 - 5)^2 + (5 - 5)^2 + (7 - 5)^2 + (8 - 5)^2]/6 = 4$.

  [Note: If the set is a "sample" rather than a "population," $\bar{x} = 5$, but $s^2 = 24/(6 - 1) = 4.8$.]

- Given a set of 25 elements with $\Sigma x = 75$ and $\Sigma x^2 = 350$, find the mean and standard deviation.
  Answer: The mean $\mu = \Sigma x/n = 75/25 = 3$.

    The variance $\sigma^2 = \Sigma x^2/n - \mu^2 = 350/25 - 3^2 = 14 - 9 = 5$.

    The standard deviation $\sigma = \sqrt{5} = 2.236$.

  [Note: If the set is a "sample," $\bar{x} = 3$, but $s^2 = (350 - 75^2/25)/(25 - 1) = 5.208$ and $s = 2.282$.]

- Suppose the attendance at a movie theater averages 780 with a standard deviation of 40. What z-score corresponds to an attendance of 835?
  Answer: $(835 - 780)/40 = 1.375$.

  What attendance corresponds to a z-score of -2.15?
  Answer: $780 - 2.15(40) = 694$.

- Estimate the standard deviation of a "bell-shaped" set of data with range 180.
  Answer: By the empirical rule σ is roughly between one-sixth and one-fourth of the range, or in this example between $180/6 = 30$ and $180/4 = 45$.

- Suppose the average noise level in a restaurant is 30 decibels with a standard deviation of 4 decibels. According to Chebyshev's theorem, what percentage of the time is the noise level between 22 and 38 decibels?
  Answer: 22 and 38 are each $8/4 = 2$ standard deviations from the mean, yielding $1 - 1/2^2 = 3/4 = 75\%$.

- Suppose $X = \{2, 9, 11, 22\}$ and $Y = \{5, 7, 15\}$. Form the set Z of differences by subtracting each element of Y from each of X:
  $Z = \{2 - 5, 2 - 7, 2 - 15, 9 - 5, 9 - 7, 9 - 15, 11 - 5, 11 - 7, 11 - 15, 22 - 5, 22 - 7, 22 - 15\} = \{-3, -5, -13, 4, 2, -6, 6, 4, -4, 17, 15, 7\}$.
  Calculate the means μ_x, μ_y, and μ_z, and the variances σ_x^2, σ_y^2, and σ_z^2. What principle is illustrated?
  Answer: $\mu_x = 11$, $\mu_y = 9$, $\mu_z = 2$, $\sigma_x^2 = 51.5$, $\sigma_y^2 = 18.67$, and $\sigma_z^2 = 70.17$. Note that $\mu_z = \mu_x - \mu_y$ and $\sigma_z^2 = \sigma_x^2 + \sigma_y^2$. This is true in general: the mean of a set of differences is equal to the *difference* of the means of the two original sets, while the variance of a set of differences is equal to the *sum* of the variances of the two original sets.

Theme 2 SHAPE

*T*here are a variety of ways to organize and arrange data. Much information can be put into tables, but these arrays of bare figures tend to be spiritless and sometimes forbidding. Some form of graphical display is often best for seeing patterns and shapes and for giving an impression of everything at once.

Key 7 Histograms

OVERVIEW *The **histogram** is an important visual representation of data in which relative frequencies are represented by relative areas.*

KEY EXAMPLE

Suppose there are 2000 families in a small town and the distribution of children among them is as follows: 300 families have no child, 400 have one child, 700 have two children, 300 have three, 100 have four, 100 have five, and 100 have six. These data can be displayed in the following *bar graph*.

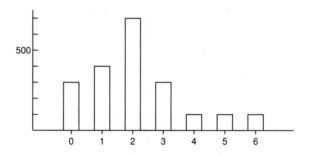

The frequencies of different results are indicated by the *heights* of the bars representing these results.

   A **histogram** can be constructed from the above bar graph by widening each bar until the sides meet at a point halfway between each two adjacent bars.

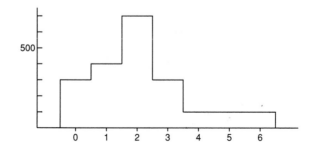

Sometimes, instead of labeling the vertical axis with *frequencies*, it is more convenient or more meaningful to use *relative frequencies*, that is, frequencies divided by the total number in the population.

Number of children	Frequency	Relative frequency
0	300	300/2000 = .150
1	400	400/2000 = .200
2	700	700/2000 = .350
3	300	300/2000 = .150
4	100	100/2000 = .050
5	100	100/2000 = .050
6	100	100/2000 = .050

Note that the shape of the histogram is the same whether the vertical axis is labeled with frequencies or with relative frequencies.

The histogram shown above indicates the relative frequency of each value from 0 to 6, but histograms can also indicate the relative frequencies of values falling between given scores.

KEY EXAMPLE

Consider the following histogram, where the vertical axis has not been labeled:

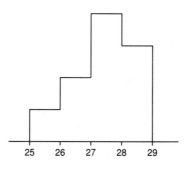

15

It is impossible to determine the actual frequencies; however, we can determine the relative frequencies by noting the fraction of the total *area* that is over any interval:

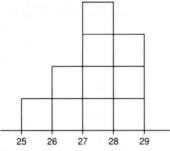

We can divide the area into ten equal portions and then note that one-tenth or 10% of the area is above 25–26, 20% is above 26–27, 40% is above 27–28, and 30% is above 28–29.

Although it will not always be possible to divide histograms so nicely into ten equal areas as above, the principle of relative frequencies corresponding to relative areas will still apply.

KEY EXAMPLE

The following histogram indicates the relative frequencies of ages of U.S. scientists in 1967.

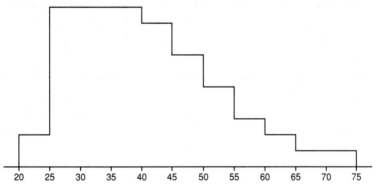

If one compares areas (or counts small rectangles!), one can conclude that 50% of the scientists are between 25 and 40 years of age, 20% are between 45 and 55 years of age, etc. Now if we also knew that there were 300,000 U.S. scientists in 1967, then we could convert these percentages to frequencies, such as 150,000 scientists between 25 and 40 years of age, or 60,000 scientists between 45 and 55 years of age.

Key 8   Histograms and measures of central tendency

OVERVIEW   *Suppose we have a detailed histogram such as the following:*

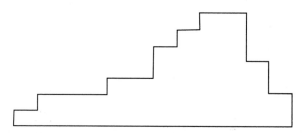

Our measures of central tendency fit naturally into such a diagram.

The *mode* is defined as the most frequent value, so it is the point or interval at which the graph is highest.

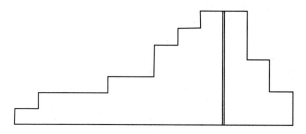

The *median* divides a distribution in half, so it is represented by a line that divides the total area of the histogram in half.

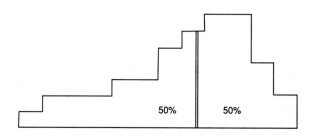

The *mean* is a value that is affected by the spacing of all the values. Therefore, if the histogram is considered to be a solid region, then the mean corresponds to a line passing through the center of gravity (where the graph would balance if it were a solid object).

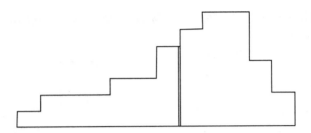

The above distribution, spread thinly far to the low side, is said to be *skewed to the left*. Note that in this case the mean is less than the median. Similarly, a distribution spread far to the high side is called *skewed to the right*, and its mean will be greater than its median.

KEY EXAMPLE

Suppose that the faculty salaries at a college have a median of $32,500 and a mean of $38,700. What does this indicate about the shape of the distribuiton of the salaries?

Answer: The median is less than the mean, so the salaries are probably skewed to the right—a few high paid professors with the bulk of the professors on the lower end of the pay scale.

Key 9   Histograms, *z*-scores, and
percentile ranks

OVERVIEW   *We have seen that relative frequencies are represented by relative areas and so labeling of the vertical axis is not crucial. If we know the standard deviation, the horizontal axis can be labeled in terms of z-scores. In fact, if we are given the percentile rankings of various z-scores, we can construct a histogram.*

KEY EXAMPLE

Suppose we are given the following data:

z-Score:	−2	−1	0	1	2
Percentile ranking:	0	20	60	70	100

We note that the entire area is less than *z*-score +2 and greater than *z*-score −2. Also, 20% of the area is between *z*-scores −2 and −1, 40% is between −1 and 0, 10% is between 0 and 1, and 30% is between 1 and 2. Thus the histogram is as follows:

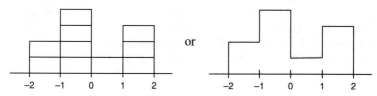

Suppose that we are also given four in-between *z*-scores:

z-Score	Percentile ranking
2.0	100
1.5	80
1.0	70
0.5	65
0.0	60
−0.5	30
−1.0	20
−1.5	5
−2.0	0

Then we have

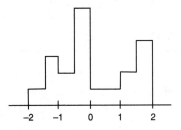

With 1000 z-scores perhaps the histogram would look like this:

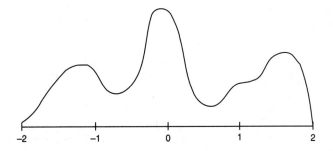

The height at any point is meaningless; what is important is relative *areas*. For example, in the final diagram, what percent of the area is between z-scores of +1 and +2?   *Answer*: Still 30%.
What percent is to the left of 0?   *Answer*: Still 60%.

Key 10 Stem and leaf displays

OVERVIEW *While a histogram may show how many scores fall into each grouping or interval, the exact values of individual scores are often lost. An alternative pictorial display, called a **stem and leaf display**, retains this individual information.*

KEY EXAMPLE

Consider the set {25, 33, 28, 31, 45, 52, 37, 31, 46, 33, 20}. Let 2, 3, 4, and 5 be place holders for 20, 30, 40, and 50. List the last digit of each value from the original set after the appropriate place holder. The result is the **stem and leaf display** of these data:

Stems	Leaves
2	5 8 0
3	3 1 7 1 3
4	5 6
5	2

Drawing a continuous line around the leaves results in a horizontal histogram:

Note that the stem and leaf display gives the shape of the histogram. However, the histogram does not indicate the values of the original data, as does the stem and leaf display.

A **revised stem and leaf display** is obtained by rearranging the digits of each row into ascending order. This ordered display shows a second level of information from the original stem and leaf picture.

KEY EXAMPLE

The revised display of the data in the above example is as follows:

2	0 5 8
3	1 1 3 3 7
4	5 6
5	2

The stems in stem and leaf displays may be other than single digits.

KEY EXAMPLE

Suppose the distribution of 25 advertised house prices (in $1000's) in a certain comunity is given by: {56, 89, 165, 73, 83, 145, 90, 189, 127, 77, 110, 112, 132, 120, 94, 130, 84, 65, 99, 154, 86, 120, 122, 103, 130}. One possible stem and leaf display of this data is as follows:

50–74	56 73 65
75–99	89 83 90 77 94 84 99 86
100–124	10 12 20 20 03
125–149	45 27 32 30 22 30
150–174	65 54
175–200	89

Note that the stems above were chosen to be intervals of length 25, and that the "1" of the 100 is left out of the leaves so that the leaves align vertically.

Key 11 Box and whisker displays

OVERVIEW *A **box and whisker display** is a visual representation of dispersion that shows the largest value, the smallest value, the median, the median of the top half of the set, and the median of the bottom half of the set.*

KEY EXAMPLE

The total farm product indexes for all years from 1919 through 1945 (with 1910–14 as 100) are as follows: 215, 210, 130, 140, 150, 150, 160, 150, 140, 150, 150, 125, 85, 70, 75, 90, 115, 120, 125, 100, 95, 100, 130, 160, 200, 200, 210. (Note the instability of prices received by farmers!) In these data, the largest value is 215, the smallest is 70, the median is 140, the median of the top half is 160, and the median of the bottom half is 100. A **box and whisker display** of these five numbers is as follows:

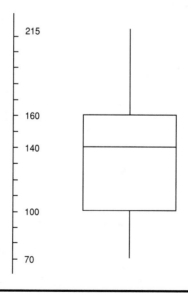

Note that the display consists of two "boxes" together with two "whiskers"—hence the name. The boxes show the spread of the two middle quarters, while the whiskers show the spread of the two outer quarters. This relatively simple display conveys information not immediately available from histograms or stem and leaf displays.

Key 12 Theme exercises with answers

OVERVIEW *Sample questions of the type that might appear on homework assignments and tests are presented with answers.*

- Suppose the 40 top-level executives of a large company receive salaries (in $1000's) distributed as follows: 1 between 20 and 30, 5 between 30 and 40, 10 between 40 and 50, 12 between 50 and 60, 6 between 60 and 70, 4 between 70 and 80, and 2 between 80 and 90. Draw a histogram with two vertical axes, one showing frequencies and the other showing relative frequencies.

Answer:

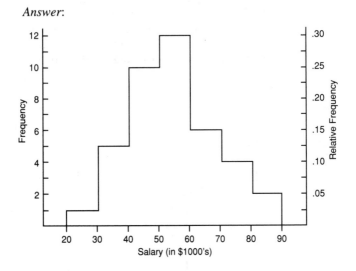

- Consider the following annual murder rates (per 100,000 people) for each of the 50 states: AL—13.3, AK—12.9, AZ—9.4, AR—9.1, CA—11.7, CO—7.3, CT—4.2, DE—6.7, FL—11.0, GA—14.4, HI—6.7, ID—5.4, IL—9.9, IN—6.2, IA—2.6, KS—5.7, KY—9.0, LA—15.8, ME—2.7, MD—8.2, MA—3.7, MI—10.6, MN—2.0, MS—12.6, MO—10.4, MT—4.8, NE—3.0, NV—15.5, NH—1.4, NM—10.2, NY—10.3, NC—10.8, ND—1.2, OH—6.9, OK—8.5, OR—5.0, PA—6.2, RI—4.0, SC—11.5, SD—1.9, TN—9.4, TX—14.2, UT—3.7, VT—3.3, VA—8.8, WA—4.6, WV—6.8, WI—2.5, WY—7.1. Draw a histogram with two horizontal axes, one corresponding to raw scores and one to *z*-scores.

Answer: Using our basic formulas gives $\mu = 7.57$ and $\sigma = 3.91$. We calculate raw scores corresponding to various z-scores by using $x = 7.57 + 3.91z$. Finally, we count elements (2% per element) to obtain percentile rankings.

z-Score:	−2	−1.5	−1	−0.5	0	1	1.5	2	2.5	3
Raw score:	−0.25	1.71	3.66	5.62	7.57	9.53	11.48	13.43	15.39	17.34
Percentile ranking:	0	4	18	36	54	68	82	92	96	100

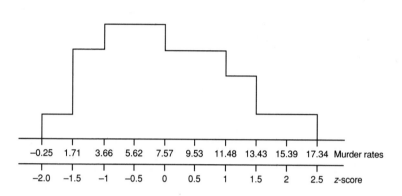

- Give a stem and leaf display for the set {126, 195, 149, 122, 189, 164, 228, 177, 165, 150, 169, 127, 176, 147, 148, 159, 128, 122, 150, 193, 207, 164, 168, 110, 155, 127, 152, 174, 190, 219, 125, 193, 141, 127, 155, 133, 150, 162, 168, 128, 125, 137, 146, 120, 154, 176, 166, 117, 154, 137} representing the 1988 per capita personal income (in $100's) for the 50 states.

Answer:

11	0 7
12	6 2 7 8 2 7 5 7 8 5 0
13	3 7 7
14	9 7 8 1 6
15	0 9 0 5 2 5 0 4 4
16	4 5 9 4 8 2 8 6
17	7 6 4 6
18	9
19	5 3 0 3
20	7
21	9
22	8

- Draw a box and whisker display of the above data.

Answer: Arranging the set in numerical order gives {110, 117, 120, 122, 122, 125, 125, 126, 127, 127, 127, 128, 128, 133, 137, 137, 141, 146, 147, 148, 149, 150, 150, 150, 152, 154, 154, 155, 155, 159, 162, 164, 164, 165, 166, 168, 168, 169, 174, 176, 176, 177, 189, 190, 193, 193, 195, 207, 219, 228}. The smallest value is 110; the largest, 228. The median is 153; the median of the top half, 169; the median of the bottom, 128.

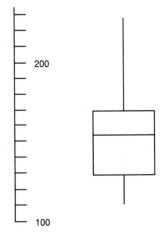

Theme 3   PROBABILITY

*I*n the world around us, sometimes unlikely events take place, and at other times very feasible events do not occur. Because of the myriad and minute origins of various happenings, it is often impracticable, or simply impossible, to predict exact outcomes. However, while we may not be able to foretell a specific result, we may be able to assign what is called a *probability* to the likelihood of any particular event happening.

For our study of statistics, we need an understanding of the probability of an elementary event that could happen many times, each time under the same circumstances. We want to be able to deduce the chance or prospect of occurrence of such events, and then use this information to make inferences about more complex circumstances where complete information is not known. For example, we might analyze the past movements of various stock prices given various economic conditions, calculate specific probabilities, and then ask what can be said, with what degree of confidence, about future movements.

In this theme, we limit our discussion to the development of the specific techniques necessary to appreciate the basic principles of statistical analysis, which will later be considered. In particular, we need to understand what are called *binomial probabilities* and be able to calculate *expected values* regarding outcomes associated with such binomial probabilities.

INDIVIDUAL KEYS IN THIS THEME

Key 13 Elementary concepts

OVERVIEW *The **probability** of an outcome of some experiment is a mathematical statement about the likelihood of that event occurring. Probabilities are always between 0 and 1, with a probability close to 0 meaning that an event is unlikely to occur, and a probability close to 1 meaning that the event is likely to occur. The sum of the probabilities of all the separate outcomes of an experiment is always 1. In this key we list basic probability concepts that are needed for our discussion of probability distributions.*

KEY EXAMPLE

Suppose A, B, C, D, and E are events, and you calculate their probabilities to be P(A) = .5, P(B) = 0, P(C) = 1, P(D) = –0.34, and P(E) = 1.6. How would you interpret these results?

Answer: Event A is as likely to happen as not to happen, event B cannot happen, and event C is sure to happen. There is some error in your calculations of the probabilities of events D and E because probabilities are never negative and are never greater than one.

Complimentary events: The probability that an event does not occur, that is, the probability of its compliment, is equal to 1 minus the probability that the event does occur.

Notation: $P(X') = 1 - P(X)$

More notation: Often p symbolizes the probability of an event and q symbolizes the probability of its compliment. Then $p + q = 1$.

KEY EXAMPLE

If the probability that a company will win a contract is .3, then the probability it will not win the contract is $1 - .3 = .7$.

Addition principle: If two events are mutually exclusive, that is, if they cannot occur simultaneously, then the probability of at least one occurring is equal to the sum of their respective probabilities. If $P(X \cap Y) = 0$, then $P(X \cup Y) = P(X) + P(Y)$ (where $X \cap Y$, read as "X intersect Y," means that both X and Y occur, while $X \cup Y$, read as "X union Y," means that either X or Y occurs or both occur).*

KEY EXAMPLE

If the probabilities of Jane, Lisa, and Mary being chosen chairwoman of the board are .5, .3, and .2, respectively, then the probability that the chairwoman will be either Jane or Mary is .5 + .2 = .7.

The above addition principle can be extended to more than two events. That is, the probability that any one of several mutually exclusive events occurs is equal to the sum of their individual probabilities.

Independence principle: If the chance of one event happening is not influenced by whether or not a second event happens, then the probability that *both* events will happen is the product of their separate probabilities.

KEY EXAMPLE

The probability that a company will receive a grant from a private concern is 1/3, while the probability that it will receive a federal grant is 1/2. If whether or not the company receives one grant is not influenced by whether or not it receives the other, then the probability of receiving both grants is $1/3 \times 1/2 = 1/6$.

The above independence principle can also be extended to more than two events. That is, given a sequence of independent events, the probability that *all* happen is equal to the product of their individual probabilities.

---

*More generally, $P(X \cup Y) = P(X) + P(Y) - P(X \cap Y)$.

Key 14 Binomial formula

OVERVIEW *In many applications there are two possible outcomes. For example, either a radio is defective or it is not defective, either the workers will go on strike or they won't go on strike, either the manager's salary is above $50,000 or it is not above $50,000. For applications in which a two-outcome situation is repeated some number of times, and the probability of each of the two outcomes remains the same for each repetition, the resulting calculations involve what are known as **binomial probabilities**.*

Notation: In the discussion below, *n*! (read as "*n* factorial") denotes the product of all the integers from *n* down to 1; that is, *n*! = $n(n-1)(n-2) \ldots (3)(2)(1)$.

KEY EXAMPLE

Suppose that 30% of the employees in a large factory are smokers. What is the probability that there will be exactly two smokers in a randomly chosen five-person work group?

Answer: We reason as follows. The probability that a person smokes is 30% = .3, so the probability that he or she doesn't smoke is 1 − .3 = .7. The probability of a particular arrangement of two smokers and three nonsmokers is $(.3)^2(.7)^3 = .03087$. The number of such arrangements is given by the *combination* $C(5,2) = 5!/2!3! = 10$. Each such arrangement has probability .03087, so the final answer is $10 \times .03087 = .3087$.

Binomial formula: Suppose an experiment has two possible outcomes, called *success* and *failure*, with the probability of success equal to *p* and the probability of failure equal to *q* (of course, *p* + *q* = 1). Suppose further that the experiment is repeated *n* times, and the outcome at any particular time does not have any influence over the outcome at any other time. Then the probability of exactly *x* successes (and thus *n* − *x* failures) is

$$C(n,x)p^x q^{n-x} = \frac{n!}{x!(n-x)!} p^x q^{n-x}$$

KEY EXAMPLE

A manager notes that there is a .125 probability that any employee will arrive late for work. What is the probability that exactly one person in a six-person department will arrive late?

Answer: If the probability of being late is .125, then the probability of being on time is $1 - .125 = .875$. If one person out of six is late, then $6 - 1 = 5$ will be on time. $C(6,1) = 6!/5!1! = 6$. Thus the desired probability is: $C(6,1)(.125)^1(.875)^5 = 6(.125)(.875)^5 = .385$.

Many, perhaps most, applications involve such phrases as *at least, at most, less than, more than*. In these cases solutions involve summing together two or more cases.

KEY EXAMPLE

A manufacturer has the following quality-control check at the end of a production line: If at least eight out of ten randomly picked articles meet all specifications, the whole shipment is approved. If, in reality, 85% of the shipment would meet all specifications, what is the probability that the shipment will make it through the control check?

Answer: The probability of meeting specifications is .85, so the probability of not meeting specifications must be .15. We want the probability that at least eight out of ten articles meet specifications, that is, the probability that exactly eight or exactly nine or exactly ten meet specifications. We sum the three binomial probabilities:

Exactly 8 out of 10 meet specifications	Exactly 9 out of 10 meet specifications	Exactly 10 out of 10 meet specifications

$$C(10,8)(.85)^8(.15)^2 + C(10,9)(.85)^9(.15)^1 + C(10,10)(.85)^{10}(.15)^0$$

$$= \frac{10!}{8!2!}(.85)^8(.15)^2 + 10(.85)^9(.15) + (.85)^{10} = .820$$

In some situations it is easier to calculate the probability of the complimentary event and subtract this value from 1.

KEY EXAMPLE

Joe DiMaggio had a career batting average of .325. What was the probability that he would get at least one hit in five official times at bat?

Answer: We could sum the probabilities of exactly one hit, two hits, three hits, four hits, and five hits. However, the complement of "at least one hit" is "zero hit." The probability of no hit is $C(5,0)(.325)^0(.675)^5 = (.675)^5 = .140$, and thus the probability of at least one hit in five times at bat is $1 - .140 = .860$.

Sometimes we are asked to calculate the probability of each of the separate outcomes (these should sum to 1).

KEY EXAMPLE

If the probability of a male birth is .51, what is the probability that a five-child family will have all boys? Exactly four boys? Exactly three boys? Exactly two boys? Exactly one boy? All girls?

Answer:

$$
\begin{aligned}
P(5 \text{ boys}) &= C(5,5)(.51)^5(.49)^0 = (.51)^5 & = .0345 \\
P(4 \text{ boys}) &= C(5,4)(.51)^4(.49)^1 = 5(.51)^4(.49) & = .1657 \\
P(3 \text{ boys}) &= C(5,3)(.51)^3(.49)^2 = 10(.51)^3(.49)^2 & = .3185 \\
P(2 \text{ boys}) &= C(5,2)(.51)^2(.49)^3 = 10(.51)^2(.49)^3 & = .3060 \\
P(1 \text{ boy}) &= C(5,1)(.51)^1(.49)^4 = 5(.51)(.49)^4 & = .1470 \\
P(0 \text{ boys}) &= C(5,0)(.51)^0(.49)^5 = (.49)^5 & = \underline{.0283} \\
& & 1.0000
\end{aligned}
$$

Note: In calculating *combinations*, 0! is defined to be 1; thus $C(5,5) = 5!/5!0! = 1$, and $C(5,0) = 5!/0!5! = 1$.

The binomial probability of x successes can be found from the probability of $x - 1$ successes using the formula:

$$
P(x \text{ successes}) = \frac{n - x + 1}{x} \frac{p}{1 - p} P(x - 1 \text{ successes})
$$

KEY EXAMPLE

A marksman can hit a bullseye target 95% of the time. Given that the probability of exactly 8 bullseyes in 10 shots is .0746, what is the probability of exactly 9 bullseyes in 10 shots?

Answer:

$$
\frac{10 - 9 + 1}{9} \frac{.95}{1 - .95} (.0746) = .315
$$

Key 15   Random variables

OVERVIEW   *Often each outcome of an experiment has not only an associated probability, but also an associated **real number**. For example, the probability might be 1/2 that there are five defective batteries; the probability might be .01 that a company will receive seven contracts; the probability might be .95 that three people will recover from a disease. If* X *represents the different numbers associated with the potential outcomes of some situation, then we call* X *a **random variable**.*

KEY EXAMPLE

A town prison official knows that 1/2 the inmates he admits stay only 1 day, 1/4 stay 2 days, 1/5 stay 3 days, and 1/20 stay 4 days before they are either released or sent on to the county jail. If X represents the number of days, then X is a *random variable* that takes the values 1, 2, 3, and 4. X takes the value 1 with probability 1/2, the value 2 with probability 1/4, the value 3 with probability 1/5, and the value 4 with probability 1/20.

The random variable in the example above is called **discrete** because it can assume only a countable number of values, while the one in the following example is called **continuous** because it can assume values associated with a whole line interval.

KEY EXAMPLE

Let X be a random variable whose values correspond to the speeds at which a jet plane can fly. Note that the jet might be traveling at 623.478 . . . mph or any other value in some whole interval. We might ask what the probability is that the plane is flying between 300 and 400 mph.

A **probability distribution** for a discrete variable is a listing or formula giving the probability for each value of the random variable.

KEY EXAMPLE

Concessionaires know that attendance at a football stadium will be 60,000 on a clear day, 45,000 if there is light snow, and 15,000 if there is heavy snow. Furthermore, the probability of clear skies, light snow, or heavy snow on any particular day is 1/2, 1/3, or 1/6 respectively. (Here we have a random variable X that takes the values 60,000, 45,000, and 15,000.)

Outcome:	Clear skies	Light snow	Heavy snow
Probability:	1/2	1/3	1/6
Random variable:	60,000	45,000	15,000

If the probabilities come from the binomial formula, then we have what is called a **binomial probability distribution**.

Key 16 Expected value or mean of a random variable

OVERVIEW *The **expected value** (or **average** or **mean**) of a random variable X (with a finite number of values) is the sum of the products obtained by multiplying each value x by the corresponding probability P(x). We write:*

$$E(X) = \sum x\, P(x)$$

KEY EXAMPLE

In a lottery, 10,000 tickets are sold with a prize of $7500 for the one winner. The actual winning payoff is $7499 because the winner paid for his $1 ticket, so we have:

Outcome:	Win	Lose
Probability:	$\dfrac{1}{10{,}000}$	$\dfrac{9999}{10{,}000}$
Random variable:	7499	−1

The expected value is $7499\left(\dfrac{1}{10{,}000}\right) + (-1)\left(\dfrac{9999}{10{,}000}\right) = -0.25.$

Thus the *average* result for each person betting the lottery is a 25¢ loss.

KEY EXAMPLE

A manager must choose among three options. Option A has a 10% chance of resulting in a $250,000 gain, but otherwise will result in a $10,000 loss. Option B has a 50% chance of gaining $40,000 and a 50% chance of losing $2000. Finally, option C has a 5% chance of gaining $800,000, but otherwise will result in a loss of $20,000. Which option should the manager choose?

Answer:

	Option A		Option B		Option C	
	Gain	Loss	Gain	Loss	Gain	Loss
Outcome:						
Probability:	.10	.90	.50	.50	.05	.95
Random variable:	250,000	−10,000	40,000	−2000	800,000	−20,000

$$E(A) = .10(250,000) + .90(-10,000) = \$16,000$$
$$E(B) = .50(40,000) \ + .50(-2000) \ \ = \$19,000$$
$$E(C) = .05(800,000) + .95(-20,000) = \$21,000$$

The manager should choose Option C!

Although option C has the greatest mean, the manager might well wish to consider the relative riskiness of each option. If, for example, a \$5000 loss would be disastrous for the company, the manager might well decide to choose option B with its maximum possible loss of \$2000. In the next key, we will consider how to measure the *variability* of a random variable.

KEY EXAMPLE

One investment has two possible returns: \$3000 with a probability 1/4 and \$2000 with probability 3/4. A second investment has possible returns of \$6000, \$7000, and \$9000 with probabilities of 1/6, 1/2, and 1/3, respectively. Assume that what happens on one investment is independent of what happens on the other. What are the expected values for the return on each investment and on the total investment?

Answer:

$$E(X) = 3000(1/4) + 2000(3/4) = \$2250$$

$$E(Y) = 6000(1/6) + 7000(1/2) + 9000(1/3) = \$7500$$

For the total investment Z we have

x	$P(x)$		x	$P(x)$
3000 + 6000	(1/4)(1/6)		9000	1/24
3000 + 7000	(1/4)(1/2)		10000	1/8
3000 + 9000	(1/4)(1/3)	or	12000	1/12
2000 + 6000	(3/4)(1/6)		8000	1/8
2000 + 7000	(3/4)(1/2)		9000	3/8
2000 + 9000	(3/4)(1/3)		11000	1/4

$$E(Z) = 9000(1/24) + \cdots + 11000(1/4) = 9750$$

Note that $E(X) + E(X) = E(Z)$.

Key 17 Variance and standard deviation
of a random variable

OVERVIEW   *Not only is the mean important, but also we would like to measure the **variability** for the values taken on by a random variable. We are dealing with chance events, so the proper tool is **variance**.*

In Key 2 variance was defined as the mean average of the squared deviations $(x - \mu)^2$. If we regard the $(x - \mu)^2$ terms as the values of some random variable (whose probability is the same as the probability of x), then the mean of this new random variable is simply $\sum(x - \mu)^2 P(x)$. This is precisely how we define the variance σ^2 of a discrete random variable:

$$\sigma^2 = \sum(x - \mu)^2 P(x)$$

KEY EXAMPLE

A highway engineer knows that his workers can lay 5 miles of highway on a clear day, 2 miles on a rainy day, and only 1 mile on a snowy day. Suppose the probabilities are as follows:

Outcome:	Clear	Rain	Snow
Probability:	.6	.3	.1
Random variable (miles of highway):	5	2	1

Then the mean, or expected value, and the variance are calculated as follows:

$$\mu = \sum x\, P(x) = 5(.6) + 2(.3) + 1(.1) = 3.7$$

$$\sigma^2 = \sum(x - \mu)^2 P(x) = (5 - 3.7)^2 (.6) + (2 - 3.7)^2 (.3) + (1 - 3.7)^2 (.1)$$
$$= 2.61$$

As before, the standard deviation σ is the square root of the variance.

A computational tool giving the same results is as follows:

$$\sigma^2 = \sum(x - \mu)^2 P(x) = \sum x^2 P(x) - \mu^2$$

KEY EXAMPLE

A particular stock investment will yield the following profit per share with the given probability:

Profit ($):	0	5	10	15	20
Probability:	.3	.3	.2	.1	.1

Then

$$\mu = 5(.3) + 10(.2) + 15(.1) + 20(.1) = 7$$

$$\sigma^2 = 25(.3) + 100(.2) + 225(.1) + 400(.1) - 49 = 41$$

$$\sigma = \sqrt{41} = 6.403$$

KEY EXAMPLE

For the last example of Key 16 on page 37, what are the three variances, and what point is illustrated?

Answer:

$$\sigma_X^2 = (3000)^2(1/4) + (2000)^2(3/4) - (2250)^2 = 187,500$$

$$\sigma_Y^2 = (6000)^2(1/6) + (7000)^2(1/2) + (9000)^2(1/3) - (7500)^2 = 1,250,000$$

$$\sigma_Z^2 = (9000)^2(1/24) + (10000)^2(1/8) + (12000)^2(1/12) + (8000)^2(1/8) + (9000)^2(3/8) + (11000)^2(1/4) - (9750)^2 = 1,437,500$$

Note that $\sigma_X^2 + \sigma_Y^2 = \sigma_Z^2$

Key 18   Mean and standard deviation

of a binomial

OVERVIEW   *In the case of a **binomial random variable**, that is, a random variable whose values are the number of "successes" in some binomial probability distribution, there is a shortcut to calculate the mean and standard deviation.*

KEY EXAMPLE

Of the automobiles produced in a particular plant, 40% have a certain defect. Suppose a company purchases five of these cars. What is the expected value for the number of cars with defects?

Answer: We might guess that the average or mean or expected value is 40% of 5 $= .4 \times 5 = 2$, but let us calculate from the definition. Letting X represent the number of cars with the defect, we have:

$$P(0) = C(5,0)(.4)^0(.6)^5 = \quad (.6)^5 \quad = .07776$$

$$P(1) = C(5,1)(.4)^1(.6)^4 = 5(.4)(.6)^4 \quad = .25920$$

$$P(2) = C(5,2)(.4)^2(.6)^3 = 10(.4)^2(.6)^3 = .34560$$

$$P(3) = C(5,3)(.4)^3(.6)^2 = 10(.4)^3(.6)^2 = .23040$$

$$P(4) = C(5,4)(.4)^4(.6)^1 = 5(.4)^4(.6) \quad = .07680$$

$$P(5) = C(5,5)(.4)^5(.6)^0 = \quad (.4)^5 \quad = .01024$$

Outcome:	0 car	1 car	2 cars	3 cars	4 cars	5 cars
Probability:	.07776	.25920	.34560	.23040	.07680	.01024
Random variable:	0	1	2	3	4	5

$$E(X) = 0(.07776) + 1(.25920) + 2(.34560) + 3(.23040)$$
$$+ 4(.07680) + 5(.01024) = 2$$

Thus, the answer turns out to be the same as would be obtained by simply multiplying the probability of "success" times the number of cases.

The following is true: If we have a binomial probability situation with the probability of success equal to p and the number of trials equal to n, then the *expected value* or *mean* number of successes for the n trials is np.

KEY EXAMPLE

An insurance salesperson is able to sell policies to 15% of the people she contacts. Suppose she contacts 120 people during a 2-week period. What is the expected value for the number of policies she sells?

Answer: We have a binomial probability with the probability of success .15 and the number of trials 120, so the mean or expected value for the number of successes is $120 \times .15 = 18$.

Similarly, for this case there is a shortcut calculation of the variance and standard deviation. We have

$$\sigma^2 = np(1 - p) \text{ and } \sigma = \sqrt{np(1 - p)}$$

KEY EXAMPLE

Of all colon cancers, 80% are cured if detected early. Among a group of 12 newly diagnosed patients, the mean and standard deviation for the number of cures are calculated:

$$\mu = np = 12(.8) = 9.6 \text{ and } \sigma = \sqrt{np(1 - p)} = \sqrt{12(.8)(.2)} = 1.39$$

Key 19 Theme exercises with answers

OVERVIEW *Sample questions of the type that might appear on homework assignments and tests are presented with answers.*

- If the probability of a defective light bulb is .1, what is the probability that exactly three out of eight light bulbs are defective?

  Answer:

  $$C(8,3)(.1)^3(.9)^5 = \frac{8!}{3!5!}(.1)^3(.9)^5 = .03306744$$

- A grocery store manager notes that 35% of the customers buying a particular product will make use of a store coupon to receive a discount. If seven people purchase the product, what is the probability that fewer than four will use a coupon?

  Answer: In this situation, "fewer than four" means zero or one or two or three.

  $$C(7,0)(.35)^0(.65)^7 + C(7,1)(.35)^1(.65)^6 + C(7,2)(.35)^2(.65)^5$$
  $$+ C(7,3)(.35)^3(.65)^4$$
  $$= (.65)^7 + 7(.35)(.65)^6 + 21(.35)^2(.65)^5 + 35(.35)^3(.65)^4$$
  $$= .800$$

- If 6% of the major success stories with regard to playing the stock market are due to illegal insider information passing, in a group of seven successful investors, what is the probability that exactly one was achieved dishonestly? That at least one was achieved dishonestly?

  Answer:

  $$P(\text{dishonest gain}) = .06 \text{ so } P(\text{honest gain}) = .94.$$

  $$P(\text{exactly one dishonest gain}) = 5(.06)(.94)^6 = .207$$

  $$P(\text{at least one dishonest gain}) = 1 - P(\text{all honest gains})$$
  $$= 1 - (.94)^7 = .352$$

- An ambulance service calculates the number of calls per day to be the following random variable:

Calls per day:	0	1	2	3	4	5	6	7	8
Probability:	.12	.15	.18	.26	.10	.08	.06	.03	.02

Calculate the expected value and variance for the calls per day variable.

Answer:

$$\mu = \Sigma xP(x)$$
$$= 0(.12) + 1(.15) + 2(.18) + 3(.26) + 4(.10) + 5(.08) + 6(.06)$$
$$\qquad + 7(.03) + 8(.02)$$
$$= 2.82$$

$$\sigma^2 = \Sigma(x - \mu)^2 P(x) = \Sigma x^2 P(x) - \mu^2$$
$$= 0(.12) + 1(.15) + 4(.18) + 9(.26) + 16(.10) + 25(.08)$$
$$\qquad + 36(.06) + 49(.03) + 64(.02) - (2.82)^2$$
$$= 3.7676$$

- Sixty percent of all new-car buyers choose automatic transmissions. For a group of five new car buyers, calculate the mean and standard deviation for the number of buyers choosing automatics.

Answer:

$$\mu = np = 5(.6) = 3.0$$

and

$$\sigma = \sqrt{np(1 - p)} = \sqrt{5(.6)(.4)} = 1.1$$

Note that these values could have been calculated in a more involved way:

$$\mu = \Sigma xP(x)$$
$$= 0[(.4)^5] + 1[5(.6)(.4)^4] + 2[10(.6)^2(.4)^3] + 3[10(.6)^3(.4)^2]$$
$$\qquad + 4[5(.6)^4(.4)] + 5[(.6)^5]$$
$$= 3.0$$

$$\sigma = \sqrt{\Sigma(x - \mu)^2 P(x)}$$
$$= \sqrt{9[.01024] + 4[.07680] + 1[.23040] + 0[.34560] + 1[.25920] + 4[.07776]}$$
$$= 1.1$$

Theme 4 PROBABILITY
DISTRIBUTIONS

*T*his theme presents several probability distributions of general interest in statistics. Such knowledge is necessary for future study and is also immediately useful as a decision-making tool for certain classes of problems.

For example, knowing the probability of finding oil, given certain geological conditions, we can use the *binomial distribution* to calculate the probability of various numbers of positive strikes for a given number of test sitings. Knowing the average number of Supreme Court vacancies during previous presidential terms, we can use the *Poisson distribution* to calculate the probability of various numbers of vacancies arising during the next 4-year term. Knowing the mean and variance of heights of U.S. Marines, we can use the *normal distribution* to calculate the probability that any Marine has a height greater than a specified value.

The binomial distribution arises from the concepts and calculations of preceding keys. The Poisson distribution can be viewed as a limiting case of the binomial when n is large and p is small. The normal distribution can be viewed as a limiting case of the binomial when p is constant but n increases without bound. Both the Poisson and the normal can be used as approximations to binomial problems. Finally, it cannot be overstressed that, even if there were very few "naturally" occurring normal distributions, the normal has tremendous value in that it describes the distribution found in a wide range of statistical experiments, investigations, and studies. This aspect will be the focus of Themes 5 and 7.

INDIVIDUAL KEYS IN THIS THEME

Key 20   Binomial distributions

OVERVIEW   *The concept of binomial probability from Key 14 can be expanded to consider tables, histograms, and the notions of mean and standard deviation.*

To review, a binomial experiment is characterized as follows: (1) the experiment consists of *n* identical trials; (2) each trial has the same two outcomes, commonly called *success* and *failure*, with probabilities *p* and $1 - p$, respectively, (3) the probabilities of success and failure remain the same from trial to trial; that is, the outcome of any one trial has no effect on the outcome of any other trial; (4) the mean is $\mu = np$; and (5) the standard deviation is $\sigma = \sqrt{np(1 - p)}$.

KEY EXAMPLE

The probability is .6 that a well driller will find water at a depth of less than 100 feet in a certain area. Wells are to be drilled for six new home owners. What is the complete probability distribution for the number of wells under 100 feet?

Answer: If $P(x)$ represents the probability of x wells under 100 feet, then

$$P(0) = \ 1(.6)^0(.4)^6 = .004$$
$$P(1) = \ 6(.6)^1(.4)^5 = .037$$
$$P(2) = 15(.6)^2(.4)^4 = .138$$
$$P(3) = 20(.6)^3(.4)^3 = .276$$
$$P(4) = 15(.6)^4(.4)^2 = .311$$
$$P(5) = \ 6(.6)^5(.4)^1 = .187$$
$$P(6) = \ 1(.6)^6(.4)^0 = \underline{.047}$$
$$1.000$$

This probability distribution has mean $\mu = np = 6(.6) = 3.6$ and standard deviation $\sigma = \sqrt{np(1 - p)} = \sqrt{6(.6)(.4)} = 1.2$. It is useful to label two horizontal axes, one with raw scores and one with *z*-scores. In this case, *z*-scores of 0, 1, and 2 correspond to 3.6, 4.8, and 6.0, while *z*-scores of −1, −2, and −3 correspond to 2.4, 1.2, and 0, respectively.

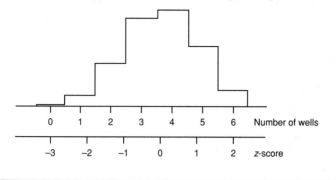

A common decision-making problem arising in manufacturing situations involves whether or not to accept a shipment of raw materials or of finished products. Concerns over quality control are sometimes handled by inspecting a sample before allowing a whole shipment to proceed. An operational rule is chosen through which the shipment will be accepted only if fewer than some specified number of defects are found in the sample. Deciding on sample size and allowable number of defects is specific to the given situation and must take into consideration time, money, quality needs, and so on. To analyze a proposed sampling plan it is important to calculate the probability of shipment acceptance, given various possible shipment defect levels.

KEY EXAMPLE

Suppose the decision-making rule is to pick a sample of size $n = 10$, and accept the whole shipment if the sample contains at most one defective item. What is the probability of acceptance if the defect level of the shipment is actually 5%? 10%? 20%? 30%? 40%? 50%?

Answer: For $p = .05$, $P(\text{accept}) = P(0 \text{ def}) + P(1 \text{ def}) = C(10,0)(.05)^0(.95)^{10} + C(10,1)(.05)^1(.95)^9 = (.95)^{10} + 10(.05)(.95)^9 = .914$. Then:

For $p = .10$, $P(\text{accept}) = (.90)^{10} + 10(.10)(.90)^9 = .736$

For $p = .20$, $P(\text{accept}) = (.80)^{10} + 10(.20)(.80)^9 = .376$

For $p = .30$, $P(\text{accept}) = (.70)^{10} + 10(.30)(.70)^9 = .149$

For $p = .40$, $P(\text{accept}) = (.60)^{10} + 10(.40)(.60)^9 = .046$

For $p = .50$, $P(\text{accept}) = (.50)^{10} + 10(.50)(.50)^9 = .011$

The graph of the above values is called the *operating characteristic curve* for this sampling plan.

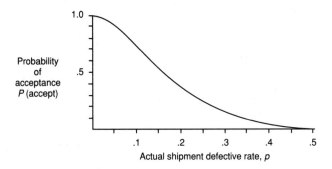

Note that, if $p = 0$, that is, there are no defective items in the entire shipment, then $P(\text{accept}) = 1$, while if $p = 1.0$, that is, the entire shipment is defective, then $P(\text{accept}) = 0$.

Key 21 Poisson distributions

OVERVIEW *The limiting case of the binomial when* n *is large and* p *is small is called the **Poisson distribution**. To calculate a Poisson probability, one needs to know only the average, μ.*

A probability distribution given by $P(x \text{ successes}) = \dfrac{\mu^x}{x!} e^{-\mu}$ where μ is the mean and $e = 2.71828\ldots$, is called a **Poisson probability distribution**.

KEY EXAMPLE

A grocery store manager knows that his store sells an average of three cans of artichoke hearts per week. Assuming that this situation is described by the formula given above, what is the probability that no can of artichoke hearts is sold in 1 week? The probability of one can in a week? Two cans in a week? One can in half a week? Three cans in 2 weeks?

Answer:

For 1 week the average is three cans, so

$$P(\text{no can in 1 week}) = e^{-3} = .050$$
$$P(\text{one can in 1 week}) = 3e^{-3} = .149$$
$$P(\text{two cans in 1 week}) = (3^2/2)e^{-3} = .224$$

For a half week the average is $0.5(3) = 1.5$, so

$$P(\text{one can in 0.5 week}) = 1.5e^{-1.5} = .335$$

For 2 weeks the average is $2(3) = 6$, so

$$P(\text{three cans in 2 weeks}) = (6^3/3!)e^{-6} = .089$$

In many examples the probabilities are multiplied by some total number to derive numerical distributions.

KEY EXAMPLE

One of the first noted examples of a Poisson distribution concerns the number of deaths from horse kicks in various corps of the German army in the late 1800s. More specifically, during the 20-year period 1875 to 1894, among the 14 cavalry corps of the German army, there was an average 0.7 deaths per corps per year as the result of horse kicks. If the deaths followed a Poisson distribution, then the descriptive probabilities per corp per year would be:

$$P(0 \text{ deaths}) = e^{-0.7} \qquad\qquad = .497$$
$$P(1 \text{ death}) = 0.7e^{-0.7} \qquad\quad = .348$$
$$P(2 \text{ deaths}) = [(0.7)^2/2]e^{-0.7} = .122$$
$$P(3 \text{ deaths}) = [(0.7)^3/3!]e^{-0.7} = .028$$
$$P(4 \text{ deaths}) = [(0.7)^4/4!]e^{-0.7} = \underline{.005}$$
$$1.000$$

If these are the correct probabilities, then the expected numerical distribution, found by multiplying the probabilities by 280 (20 years × 14 corps = 280 groups), is as follows:

Deaths	Predicted number of groups
0	.497 × 280 = 139.2
1	.348 × 280 = 97.4
2	.122 × 280 = 34.2
3	.028 × 280 = 7.8
4	.005 × 280 = 1.4
	280.0

How accurate is this description? Here are the actual figures:

Deaths	Actual number of groups
0	144
1	91
2	32
3	11
4	2

To calculate a binomial distribution, we must be given n, the number of cases under consideration. In other words, to find the probability of a certain number of successes, we must know the corresponding number of failures. However, there are situations in which the number of failures is impossible to determine. For example, in determining the probability that a hospital will have two appendectomy cases in one day, it makes little sense to ask how many appendicitis cases will not occur. In calculating the probability that a baseball team will score five runs in a game, it is meaningless to ask how many runs will not be scored. And in finding the probability that there will be ten incoming telephone calls at a switchboard, it is impossible to say how many calls will not come in. Fortunately there are many cases in which n is very large, p is very small, and the mean or average, $\mu = pn$ is both moderate and known. Then the Poisson distribution can be applied as above.

KEY EXAMPLE

A large number n of planes fly into a major airport, and the probability p that any particular plane is on the runway at any particular moment in time is small. However, the average number np of planes waiting in line on the runway at noon each day may be determined without evaluating either p or n. Once $\mu = np$ is calculated, then the Poisson formula may be applied (see Key 22 for information as to when this is appropriate).

The Poisson is used in certain types of hypothesis tests.

KEY EXAMPLE

Observations taken before a company began to dump pollutants into a river indicated that an averge of three trout per hour swim past the dump site. An inspector plans to spend an hour at the site, and will issue a warning if she sees less than three trout. If the pollutants have no effect on the trout, and so the mean per hour is still three, what is the probability that there will be a warning?

Answer:

$$P(\text{less than 3}) = P(0) + P(1) + P(2) = e^{-3} + 3e^{-3} + [3^2/2]e^{-3} = .423$$

If the pollutants really kill half of the trout, what is the probability that a warning does or does not result?

Answer: In this case, the mean is .5(3) = 1.5, and thus we have

$$P(\text{warning}) = e^{-1.5} + 1.5e^{-1.5} + [(1.5)^2/2]e^{-1.5} = .809$$

$$P(\text{no warning}) = 1 - .809 = .191$$

Note that with this inspection procedure, there still might be a warning (.423 probability) even if the pollutant is harmless, and there might be no warning (.191 probability) even if the pollutant kills half the fish.

Key 22 Poisson approximation to the binomial

OVERVIEW *An important use of the Poisson is as an approximation to the binomial. The Poisson approximation is of course valuable when* p *and* n *are not known, but the mean* np *is known. However, the Poisson approximation is useful even when* p *and* n *are known, because the calculation of a Poisson probability is simpler than the calculation of the corresponding binomial probability.*

A rough rule of thumb is that the resulting approximation is "close" provided that $n \geq 20$ and $p \leq .05$. Of course, how close is close enough will depend on the degree of accuracy required in the particular application.

KEY EXAMPLE

In a binomial distribution with $p = .02$ and $n = 70$, what is the actual probability of three successes, and what is the Poisson approximation?

Answer: We calculate this probability in two ways:

Binomial: Since the probability of success is .02, the probability of failure is .98, so

$$P(3 \text{ successes}) = \frac{70!}{67!3!}(.02)^3(.98)^{67} = .1131$$

Poisson: The mean is $70(.02) = 1.4$, so

$$P(3 \text{ successes}) = \frac{(1.4)^3}{3!}e^{-1.4} = .1128$$

Note how much easier the Poisson calculation is in comparison to the binomial calculation.

KEY EXAMPLE

A delivery of ten items is received from a manufacturing plant at which 5% of the items produced are defective. What is the probability of no defective items among the ten received? Of exactly one defective?

Answer: Since $n = 10$ is not large, we do not expect to obtain a very good approximation to the normal by using the Poisson (however, the result is surprisingly close). The mean is $10(.05) = 0.5$.

P(no defective) $= e^{-0.5}$ $= .607$ (Actual answer is .599.)
P(one defective) $= 0.5e^{-0.5} = .303$ (Actual answer is .315.)

KEY EXAMPLE

There were 231 deaths at an upstate New York hospital during a recent leap year. This is a binomial, but n is large and unknown while p is small and unknown, so we use the Poisson as an approximation. Mathematically, how many days would we have expected to have no deaths? One death? Two deaths? Three deaths? Four deaths?

Answer: $\mu = 231/336 = 0.631$ deaths per day, and thus

Outcome	Probability	Expected occurrences
0 deaths	$e^{-0.631} = .532$	$.532 \times 366 = 195$
1 death	$0.631e^{-0.631} = .366$	$.336 \times 366 = 123$
2 deaths	$(0.631^2/2!)e^{-0.631} = .106$	$.106 \times 366 = 39$
3 deaths	$(0.631^3/3!)e^{-0.631} = .022$	$.022 \times 366 = 8$
4 deaths	$(0.631^4/4!)e^{-0.631} = .004$	$.004 \times 366 = 1.5$

(The actual numbers were surprisingly close: 201 days with no deaths, 118 days with one death, 33 days with two deaths, 10 days with three deaths, 3 days with four deaths, and 1 day with five deaths.)

Key 23 Normal distributions

OVERVIEW *The normal curve is bell shaped and symmetrical with an infinite base. There are long, flattened tails that cover many values but only a small proportion of the area. The mean here is the same as the median and is located at the center. It is convenient to measure distances under the normal curve in terms of z-scores (fractions or multiples of standard deviations from the mean).*

Table A in the Appendix gives proportionate areas under the normal curve. Because of the symmetry of the curve, it is sufficient to give areas from the mean to positive z values. Table A shows, for example, that between the mean and a z-score of 1.2 there is .3849 of the area:

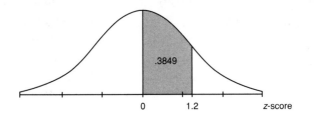

while .4834 of the area is between the mean and a z-score of −2.13.

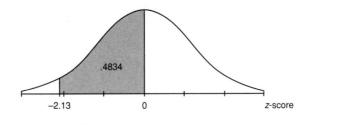

KEY EXAMPLE

The life expectancy of a particular brand of light bulbs is normally distributed with a mean of 1500 hours and a standard deviation of 75 hours.

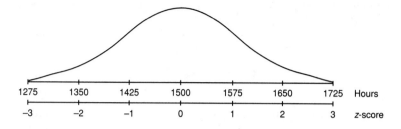

What is the probability that a bulb will last between 1500 and 1650 hours?

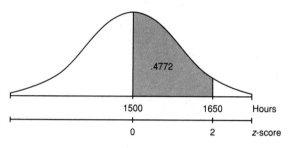

Answer: The *z*-score of 1500 is 0, the *z*-score of 1650 is $(1650 - 1500)/75$ = 2, and 2 in Table A gives a probability of .4772.

What percentage of the light bulbs will last between 1485 and 1500 hours?

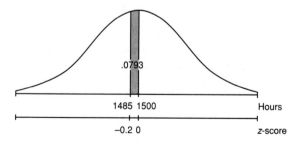

Answer: The *z*-score of 1485 is $(1485 - 1500)/75 = -0.2$, and 0.2 in Table A gives a probability of .0793 or 7.93%.

What is the probability that a bulb will last between 1416 and 1677 hours?

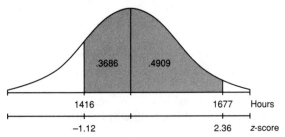

Answer: The *z*-score of 1416 is (1416 − 1500)/75 = −1.12, and the *z*-score of 1677 is (1677 − 1500)/75 = 2.36. In Table A, 1.12 gives a probability of .3686, and 2.36 gives a probability of .4909. The total probability is .3686 + .4909 = .8595.

What is the probability that a light bulb will last between 1563 and 1648 hours?

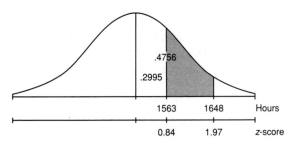

Answer: The *z*-score of 1563 is (1563 − 1500)/75 = 0.84, and, the *z*-score of 1648 is (1648 − 1500)/75 = 1.97. In Table A, 0.84 and 1.97 give probabilities of .2995 and .4756. Between 1563 and 1648 there is a probability of .4756 − .2995 = .1761.

What is the probability that a light bulb will last less than 1410 hours?

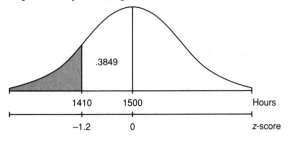

Answer: The *z*-score of 1410 is (1410 – 1500)/75 = -1.2. In Table A, 1.2 gives a probability of .3849. The probability of being less than 1500 is .5, and so the probability of being less than 1410 is .5 – .3849 = .1151.

KEY EXAMPLE

A packing machine is set to fill a cardboard box with a mean average of 16.1 ounces of cereal. Suppose the amounts per box form a normal distribution with standard deviation equal to 0.04 ounce.

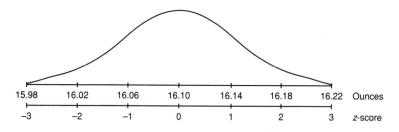

What percent of the boxes end up with at least 1 pound of cereal?

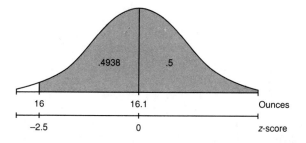

Answer: The *z*-score of 16 is (16 – 16.1)/0.04 = –2.5, and 2.5 in Table A gives a probability of .4938. The probability of more than 1 pound is .4938 + .5 = .9938 or 99.38%.

Ten percent of the boxes will contain above what number of ounces?

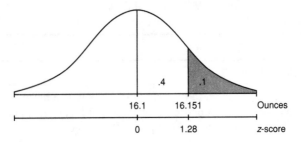

Answer: In Table A, we note that .4 area (actually .3977) is found between the mean and a *z*-score of 1.28, so to the right of a 1.28 *z*-score must be 10% of the area. Converting the *z*-score of 1.28 into a raw score yields 16.1 + 1.28(0.04) = 16.151 ounces.

Eighty percent of the boxes will contain above what number of ounces?

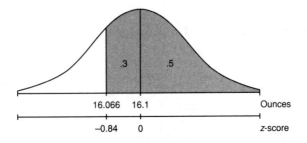

Answer: In Table A, we note that .3 area (actually .2995) is found between the mean and a *z*-score of 0.84, so to the right of a –0.84 *z*-score must be 80% of the area. Converting the *z*-score of –0.84 into a raw score yields 16.1 – 0.84(0.04) = 16.066 ounces.

Key 24   Commonly used probabilities

and their *z*-scores

OVERVIEW   *There is often an interest in the limits enclosing some specified middle percentage of the data and in values with particular percentile rankings.*

For future reference, we note, in terms of *z*-scores, the limits most often asked for.

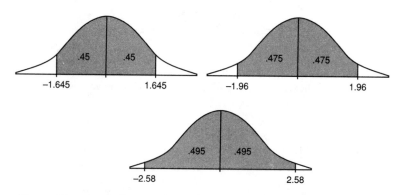

Since .45 + .45 = .90, then 90% of the values are between *z*-scores of −1.645 and +1.645. Since .475 + .475 = .95, then 95% of the values are between *z*-scores of −1.96 and +1.96. Since .495 + .495 = .99, then 99% of the values are between *z*-scores of −2.58 and +2.58.

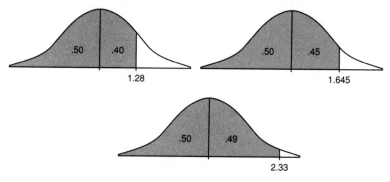

Thus, 90% of the values are below a z-score of 1.28, 95% of the values are below a z-score of 1.645, and 99% of the z-scores are below a z-score of 2.33.

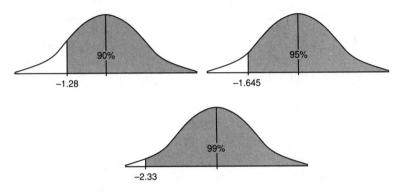

Similarly, 90% of the values are above a z-score of −1.28, 95% of the values are above a z-score of −1.645, and 99% of the values are above a z-score of −2.33.

It is also useful to note the percentages corresponding to values falling between integer z-scores.

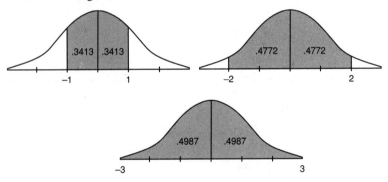

For example, .3414 + .3413 = .6826, and so 68.26% of the values are between z-scores of −1 and +1. Also, .4772 + .4772 = .9544, and so 95.44% of the values are between z-scores of −2 and +2. And .4987 + .4987 = .9974, and so 99.74% of the values are between z-scores of −3 and +3.

KEY EXAMPLE

Suppose that the average height of adult males in a particular locality is 70 inches with a standard deviation of 2.5 inches. If the distribution is normal, then the middle 95% of males are between what two heights?

Answer: As noted above, the critical z-scores are ±1.96, so the two limiting heights are 1.96 standard deviations from the mean. Therefore, $70 \pm 1.96(2.5) = 70 \pm 4.9$ or from 65.1 to 74.9 inches.

Ninety percent of the heights are below what value?

Answer: The critical z-score is 1.28, and so the height in question is $70 + 1.28(2.5) = 70 + 3.2 = 73.2$ inches.

Ninety-nine percent of the heights are above what value?

Answer: The critical z-score is −2.33, so the height in question is $70 - 2.33(2.5) = 70 - 5.825 = 64.175$ inches.

What percentage of the heights are between z-scores of ±1? Of ±2? Of ±3?

Answer: 68.26%, 95.44%, and 99.74%, respectively.

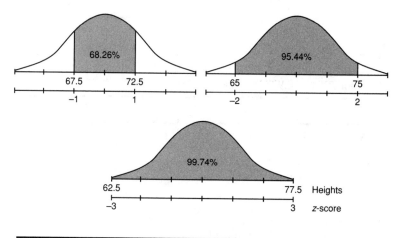

Key 25 Finding means and standard deviations in normal distributions

OVERVIEW *Knowing that a distribution is normal allows for calculations of the mean μ and the standard deviation σ, using percentage information from the population.*

KEY EXAMPLE

Given a normal distribution with a mean of 25, what is the standard deviation if 18% of the values are above 29?

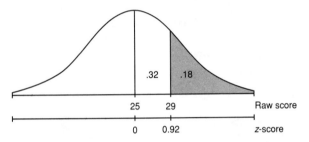

Answer: Looking for a .32 probability in Table A, we note that the corresponding *z*-score is 0.92. Thus 29 – 25 = 4 is equal to 0.92 standard deviation, that is, 0.92σ = 4, and σ = 4/0.92 = 4.35.

KEY EXAMPLE

Given a normal distribution with a standard deviation of 10, what is the mean if 21% of the values are below 50?

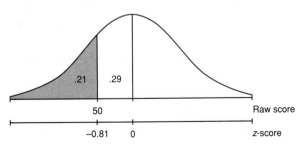

Answer: Looking for a .29 probability in Table A leads to a *z*-score of −0.81. Thus 50 is −0.81 standard deviation from the mean, and so μ = 50 + 0.81(10) = 58.1.

KEY EXAMPLE

Given a normal distribution with 80% of the values above 125 and 90% of the values above 110, what are the mean and the standard deviation?

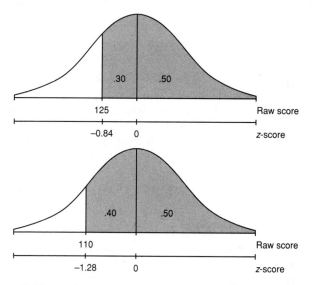

Answer: Table A gives critical *z*-scores of −0.84 and −1.28. Thus we have $(125 - μ)/σ = -0.84$ and $(110 - μ)/σ = -1.28$. Solving the system $\{125 - μ = -0.84σ,\ 110 - μ = -1.28σ\}$ simultaneously gives μ = 153.64 and σ = 34.09.

Key 26   Normal approximation to the
binomial

OVERVIEW   *Many practical applications of the binomial involve examples where* n *is large. However, for large* n, *binomial probabilities can be quite messy to calculate. Since the normal can be viewed as a limiting case of the binomial, it is natural to use the normal to approximate the binomial in appropriate situations.*

The binomial takes values only at integers, while the normal is continuous with probabilities corresponding to areas over intervals. Therefore, we must set down some technique for converting from one distribution to the other. For approximation purposes we do as follows. Each binomial probability will correspond to the normal probability over a unit interval centered at the desired value. Thus, for example, to approximate the binomial probability of eight successes we determine the normal probability of being between 7.5 and 8.5.

KEY EXAMPLE

Suppose that 15% of the cars coming out of an assembly plant have some defect. In a delivery of 40 cars what is the probability that exactly 5 cars have defects?

Answer: The actual answer is $(40!/35!5!)(.15)^5(.85)^{35}$ but as can be seen this involves a nontrivial calculation. To approximate the answer using the normal, we first calculate the mean μ and the standard deviation σ as follows:

$$\mu = np = 40(.15) = 6$$
$$\sigma = \sqrt{np(1 - p)} = \sqrt{40(.15)(.85)} = 2.258$$

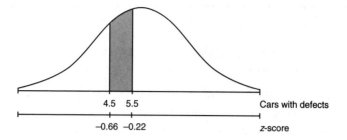

4.5　5.5　　　　　　　Cars with defects

−0.66 −0.22　　　　　　z-score

We then calculate the appropriate z-scores: $(4.5 - 6)/2.258 = -0.66$ and $(5.5 - 6)/2.258 = -0.22$. Looking up the corresponding probabilities in Table A gives a final answer of $.2454 - .0871 = .1583$. (The actual answer is $.1692$.)

Even more useful are approximations relating to probabilities over intervals.

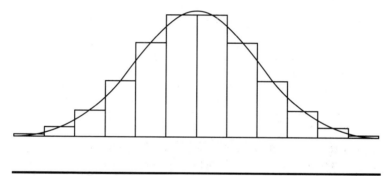

KEY EXAMPLE

If 60% of the population support massive federal budget cuts, what is the probability that in a survey of 250 people at most 155 people support the cuts?

Answer: The actual answer is the sum of 156 binomial expressions:

$$(.4)^{250} + \cdots + \frac{250!}{155!95!}(.6)^{155}(.4)^{95}$$

However, a good approximation can be obtained quickly and easily using the normal. We calculate μ and σ:

$$\mu = np = 250(.6) = 150$$
$$\sigma = \sqrt{np(1 - p)} = \sqrt{250(.6)(.4)} = 7.746$$

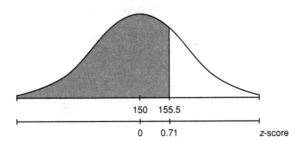

The binomial of at most 155 successes corresponds to the normal probability of ≤155.5. The z-score of 155.5 is $(155.5 - 150)/7.746 = 0.71$, and using Table A leads to a final answer of $.5000 + .2611 = .7611$.

Is the normal a good approximation? The answer, of course, depends on the error tolerances in particular situations. However, a general rule of thumb is that the normal is a "good" approximation to the binomial whenever np and $n(1 - p)$ are both greater than 5.

KEY EXAMPLE

A form of cancer is fatal in 30% of all diagnosed cases. A new drug is tried on 200 patients with this disease, and researchers will judge the new medication effective if at least 150 of the patients recover. If the medication has no effect, what is the probability that at least 150 patients recover?

Answer: The actual answer is the sum of 51 binomial expressions:

$$\frac{200!}{150!50!}(.7)^{150}(.3)^{50} + \frac{200!}{151!49!}(.7)^{151}(.3)^{49} + \cdots + (.7)^{200}$$

that would be very tedious to calculate. However, an approximate answer using the normal is readily calculated.

$$\mu = np = 200(.7) = 140$$
$$\sigma = \sqrt{np(1 - p)} = \sqrt{200(.7)(.3)} = 6.48$$

The binomial probability of 150 successes corresponds to the normal probability on the interval from 149.5 to 150.5, and thus the binomial probability of at least 150 successes corresponds to the normal probability of ≥ 149.5. The z-score of 149.5 is $(149.5 - 140)/6.48 = 1.47$. Using the normal probability table gives a final answer of $5.000 - .4292 = .0708$.

Key 27 Theme exercises with answers

OVERVIEW *Sample questions of the type that might appear on homework assignments and test are presented with answers.*

- A baseball player, with a batting average of .250, has 12 official at-bats in a three-game series. What is the probability distribution for the number of hits he makes? Display in a histogram.

  Answer: The probability of a hit is .250, and we have the following:

  $$P(0) = C(12,0)(.250)^0(.750)^{12} = 1(.75)^{12} \qquad = .032$$
  $$P(1) = C(12,1)(.250)^1(.750)^{11} = 12(.25)(.75)^{11} = .127$$
  $$P(2) = C(12,2)(.250)^2(.750)^{10} = 66(.25)^2(.75)^{10} = .232$$
  $$P(3) = C(12,3)(.250)^3(.750)^9 = 220(.25)^3(.75)^9 = .258$$
  $$P(4) = C(12,4)(.250)^4(.750)^8 = 495(.25)^4(.75)^8 = .194$$
  $$P(5) = C(12,5)(.250)^5(.750)^7 = 792(.25)^5(.75)^7 = .103$$
  $$P(6) = C(12,6)(.250)^6(.750)^6 = 924(.25)^6(.75)^6 = .040$$
  $$P(7) = C(12,7)(.250)^7(.750)^5 = 792(.25)^7(.75)^5 = .012$$
  $$P(8) = C(12,8)(.250)^8(.750)^4 = 495(.25)^8(.75)^4 = .002$$

  To three decimal places, $P(9) = P(10) = P(11) = P(12) = 0$.

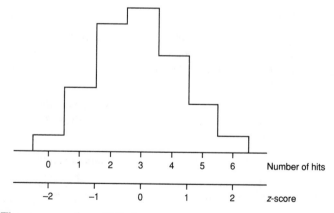

  The mean number of hits is

  $$\mu = np = 12(.25) = 3$$

and the standard deviation is

$$\sigma = \sqrt{np(1-p)} = \sqrt{12(.25)(.75)} = 1.5$$

Thus, z-scores of 0, 1, and 2 correspond to 3, 4.5, and 6, while z-scores of –1 and –2 correspond to 1.5 and 0, respectively.

- In a large northeastern town the average number of business bankruptcies per week is 1.5. What is the probability of no bankruptcy in a week? Of one bankruptcy? Of two bankruptcies? Of three?

  Answer: There are a large number of businesses, the probability of bankruptcy is small, and the average is known; therefore, we apply the Poisson distribution to obtain:

  $$P(\text{no bankruptcy}) = e^{-1.5} \qquad = .223$$
  $$P(1 \text{ bankruptcy}) = 1.5e^{-1.5} \qquad = .335$$
  $$P(2 \text{ bankruptcies}) = [(1.5)^2/2]e^{-1.5} = .251$$
  $$P(3 \text{ bankruptcies}) = [(1.5)^3/3!]e^{-1.5} = .126$$

  What is the probability of no bankruptcy in a 2-week period?

  Answer: The average for 2 weeks is 2(1.5) = 3, so

  $$P(0 \text{ bankruptcy in 2 weeks}) = e^{-3} = .0498$$

  What is the probability of four bankruptcies in a 3-week period?

  Answer: The average for 3 weeks is 3(1.5) = 4.5, so

  $$P(4 \text{ bankruptcies in 3 weeks}) = [(4.5)^4/4!]e^{-4.5} = .1898$$

- The resistors in a shipment have an average resistance of 200 ohms with a standard deviation of 5 ohms. If a normal distribution is assumed, what percentage of the resistors have resistances between 195 and 212 ohms?

  Answer: The z-scores of 195 and 212 are (195 – 200)/5 = –1 and (212 – 200)/5 = 2.4, respectively. Table A gives .3413 + .4918 = .8331 = 83.31%.

  Thirty percent of the resistors have resistances below what value?

  Answer: For a probability of .5 – .3 = .2, Table A gives a z-score of –0.52. Converting to a raw score gives 200 – 0.52(5) = 197.4 ohms.

  The middle 95% have resistances between what two values?

  Answer: The critical z-scores ±1.96 convert to 200 ± 1.96(5) = 200 ± 9.8 or between 190.2 and 209.8 ohms.

- Airline companies know that 4% of all reservations received will be no-shows, so they overbook accordingly. Suppose that there are 126 seats on a plane, and the airline books 130 reservations. What is the probability that more than 126 confirmed passengers will show up? In other words, what is the probability that the number of no-shows will be 3 or less? Solve using normal and Poisson approximations to the binomial.

Answer: Both $np = 130(.04) = 5.2$ and $n(1 - p) = 130(.96) = 124.8$ are greater than 5, so it is reasonable to determine a normal approximation. Also, both $p = .04 \leq .05$ and $n = 130 \geq 20$, so it is reasonable to calculate a Poisson approximation.

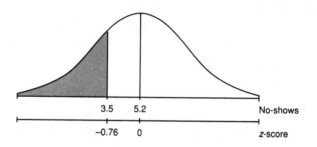

The normal approximation is calculated using $n = 130$ and $p = .04$ to derive $\mu = 5.2$ and $\sigma = 2.2343$. The z-score of 3.5 is $(3.5 - 5.2)/2.2343 = -0.76$, and the probability approximation is $.5 - .2764 = .2236$. The Poisson approximation to the binomial is calculated as follows:

$$\frac{(5.2)^3}{3!} e^{-5.2} + \frac{(5.2)^2}{2} e^{-5.2} + (5.2)e^{-5.2} + e^{-5.2} = .2381$$

The actual answer is the sum of four binomial probabilities:

$$\frac{130!}{127!3!}(.04)^3(.96)^{127} + \frac{130!}{128!2!}(.04)^2(.96)^{128} + 130(.04)(.96)^{129}$$
$$+ (.96)^{130} = .2323$$

Theme 5 THE POPULATION MEAN

*T*he point of view of statistics is illustrated by the following example. Suppose a jar is filled with black and white balls. If we knew the exact number of balls of each color, we could apply probability theory to predict the likelihood of drawing a ball of a particular color or of drawing a sample of a particular mixture. On the other hand, suppose we do not know the composition of the whole jar, but we have drawn a sample. Statistics tells us with what confidence we can estimate the composition of the whole jar from that of the sample.

In our daily lives we frequently make judgments on the basis of observing the characteristics of samples. A consumer looks at a few sales slips and concludes that a clothing shop is a high- or a low-priced store. A high school senior talks with a group of students and then reaches an opinion about the college these students represent. A traveling salesman rings doorbells for a day and then makes an estimate as to the profitability of a new territory. In all these examples complete observations are impossible or at least impracticable. The same is true in more structured situations. A medical researcher can't test everyone who has a particular form of cancer, a retailer can't test every flashbulb from a shipment (there would be none left to sell!), and a pollster can't survey every potential voter.

Thus we use samples to make inferences about characteristics of the whole population. Many questions arise. With what size and in what manner should a sample be chosen? What conclusions about the population can be drawn from the sample? With what degree of confidence can these conclusions be stated? In answering these questions we keep in mind that we are usually considering only a few members out of a large population and thus can never make any inference about the whole population with 100% certainty. We can, however, draw inferences with specified degrees of certainty.

Key 28   The distribution of sample means

OVERVIEW   *Heights, weights, and the like tend to result in normal distributions, but normally distributed natural phenomena occur much less frequently than one might guess. However, the normal curve has an importance in statistics that is independent of whether or not it appears in nature. What is significant is that the results of many types of sampling experiments can be analyzed using the normal curve. For example, there is no reason to suppose that the amounts of money that different people spend in grocery stores are normally distributed. However, if every day we survey 30 people leaving a store and determine the average grocery bill, then these daily averages will have a nearly normal distribution. This statistical key will enable us to use a sample mean to estimate a population mean.*

We are interested in estimating the mean of a population. For our estimate we could simply randomly pick a single element of the population, but we then would have little confidence in our answer. Suppose instead we pick 100 elements and calculate their average. It is intuitively clear that the resulting sample mean has a greater chance of being closer to the mean of the whole population than does the value for any individual member of the population.

When we pick a sample and measure its mean, we are finding exactly one sample mean out of a whole universe of sample means. To judge the significance of a single sample mean, we must know how sample means vary. Consider the set of means from all possible samples of a specified size. It is both apparent and reasonable that the sample means will be clustered about the mean of the whole population; furthermore, these sample means will have a tighter clustering than do the elements of the original population. In fact, we might guess that the larger the chosen sample size, the tighter will be the clustering.

The following principle forms the basis of much of what we do in following keys. It is a simplified statement of the *central limit theorem* of statistics.

Start with a population with a given mean μ and standard deviation σ. Pick n sufficiently large (at least 30), and take all samples of size n. Compute the mean of each of these samples. Then:

- The set of all sample means will be approximately *normally* distributed.

- The *mean* of the set of sample means will equal μ, the mean of the population.

- The *standard deviation*, $\sigma_{\bar{x}}$, of the set of sample means will be approximately equal to $\sigma / \sqrt{n}$, that is, equal to the standard deviation of the whole population divided by the square root of the sample size.

KEY EXAMPLE

Suppose that the average outstanding credit card balance among young couples is \$650 with a standard deviation of \$420. If 100 couples are selected at random, what is the probability that the mean outstanding credit card balance exceeds \$700?

Answer: The sample size is over 30, so by the central limit theorem, the set of sample means is approximately normally distributed with mean 650 and standard deviation $420 / \sqrt{100}$ = 42. With a z-score of $(700 - 650)/42 = 1.19$, the probability of our sample mean exceeds 700 is $.5000 - .3830 = .1170$.

Key 29 Confidence interval estimate
of the mean

OVERVIEW *Using a measurement from a sample, we will never be able to say exactly what the population mean is; rather, we will always say we have a certain confidence that the population mean lies in a certain interval.*

The central limit theorem gives us the probability that a sample mean lies within a specified interval around the population mean, but this is precisely the same as saying that the population mean lies within a specified interval around the sample mean.

KEY EXAMPLE

A bottling machine is operating with a standard deviation of 0.12 ounce. Suppose that in a sample of 36 containers the machine dumped an average of 16.1 ounces. Estimate the mean number of ounces in all bottles that this machine fills. More specifically, give an interval in which we are 95% certain that the mean lies.

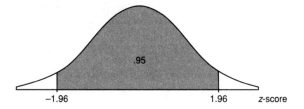

Answer: For samples of size 36, the sample means are approximately normally distributed with a standard deviation of

$$\sigma_{\bar{x}} = \sigma / \sqrt{n} = 0.12 / \sqrt{36} = 0.02$$

From Key 24 we have that 95% of the sample means should be within 1.96 standard deviations of the population mean. Equivalently, we are 95% certain that the population mean is within 1.96 standard deviations of any sample mean.* In our case, $16.1 \pm 1.96(0.02) = 16.1 \pm 0.0392$, and we are 95% sure that the mean number of ounces in all bottles is between 16.0608 and 16.1392. This is called a *95% confidence interval estimate*.

---

*Note that we cannot say there is a .95 *probability* that the population mean is within 1.96 standard deviations of a given sample mean. For a given sample mean, the population mean either is or isn't within the specified interval, so the probability is either 1 or 0.

How about a 99% confidence interval estimate?

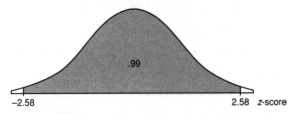

-2.58   2.58   z-score

Answer: Here, $16.1 \pm 2.58(0.02) = 16.1 \pm 0.0516$, and we are 99% sure that the mean number of ounces in all bottles is between 16.0484 and 16.1516.

Note that, when we wanted a higher certainty (99% instead of 95%), we had to settle for a larger, less specific interval (± 0.0516 instead of ± 0.0392).

Frequently we do not know σ, the population standard deviation. In these cases, we must use s, the standard deviation of the sample, as an estimate for σ. We calculate the standard deviation of a sample by the formula

$$s = \sqrt{\frac{\sum(x - \bar{x})^2}{n-1}}$$

whenever we intend to use s as an estimate for σ.

KEY EXAMPLE

An advertiser wishes to determine the mean number of hours per week that teenagers spend before television sets. The results of 500 interviews are as follows: $\sum x = 16{,}475$ and $\sum(x - \bar{x})^2 = 48{,}907$. Determine a 98% confidence interval estimate.

Answer: We first calculate the sample mean and standard deviation:

$$\bar{x} = \frac{\sum x}{n} = \frac{16475}{500} = 32.95, \quad s = \sqrt{\frac{\sum(x - \bar{x})^2}{n-1}} = \sqrt{\frac{48907}{499}} = 9.9$$

We use $s = 9.9$ as an estimate for the population standard deviation σ to find the standard deviation of the sample means,

$$\sigma_{\bar{x}} = \sigma / \sqrt{n} = 9.9 / \sqrt{500} = 0.443$$

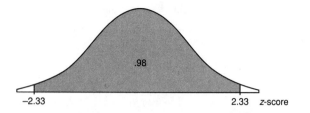

We can be 98% sure that the mean number of hours per week that teenagers spend watching television is in the range $32.95 \pm 2.33(0.443) = 32.95 \pm 1.03$, or between 31.92 and 33.98 hours per week.

KEY EXAMPLE

A new drug results in lowering heart rates by varying amounts with a standard deviation of 2.49 beats power minute. Find a 95% confidence interval estimate for the mean lowering of heart beats in all patients if a 50 person sample averages a drop of 5.32 beats per minute.

Answer: The standard deviation of sample means is

$$\sigma_{\bar{x}} = \sigma/\sqrt{n} = 2.49/\sqrt{50} = 0.352$$

We are 95% certain that the mean lowering of heart beats is in the range $5.32 \pm 1.96(0.352) = 5.32 \pm 0.69$ or between 4.63 and 6.01 heartbeats per minute.

With what certainty can we assert that the new drug lowers heart rates by a mean of 5.32 ± 0.75 beats per minute?

Answer: Converting 0.75 to a z-score yields $0.75/0.352 = 2.13$. A z-score of 2.13 gives .4834 in Table A and our answer is $2(.4834) = .9668$. That is, 5.32 ± 0.75 beats per minute is a 96.68% confidence interval estimate of the mean lowering of heart rate effected by this drug.

Key 30   Selecting a sample size

OVERVIEW   *Statistical principles are useful not only in analyzing data, but also in setting up experiments. One consideration is the choice of a sample size. In making interval estimates of population means, we have seen that each inference goes hand in hand with an associated confidence-level statement. Generally, if we want a smaller, more precise interval estimate, we either decrease the degree of confidence or increase the sample size. Similarly, if we want to increase the degree of confidence, we may either accept a wider interval estimate or increase the sample size. Choosing a larger sample size seems always desirable; in the real world, however, this decision involves time and cost considerations.*

KEY EXAMPLE

Ball bearings are manufactured by a process that results in a standard deviation in diameter of 0.025 inch. What sample size should be chosen if one wishes to be 99% sure of knowing the diameter to within ±0.01 inch?

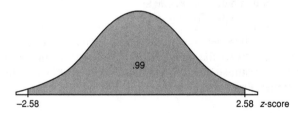

Answer: We have $\sigma_{\bar{x}} = \sigma / \sqrt{n} = 0.025 / \sqrt{n}$ and $2.58\sigma_{\bar{x}} \leq 0.01$. Thus $2.58(0.025 / \sqrt{n}) \leq 0.01$. Algebraically we find that $\sqrt{n} \geq 2.58(0.025)/0.01 = 6.45$, so $n \geq 41.6$. We choose a sample size of 42.

KEY EXAMPLE

A sociologist is designing an experiment to determine the mean age of U.S. citizens who have strong opinions against committing funds for military operations in Central America. She has determined that, for a 90% confidence estimate of the mean age within ±3.0 years, she will need to survey 100 individuals having the specified opinions. What would be the sample size for a 90% confidence estimate of the mean age within ±1.5 years? Within ±1.0 year?

Answer: We have $\sigma_{\bar{x}} = \sigma / \sqrt{100}$ and $1.645\sigma_{\bar{x}} \leq 3.0$, or $1.645(\sigma / \sqrt{100}) \leq 3.0$. We want to know n_1 and n_2, where $1.645(\sigma / \sqrt{n_1}) \leq 1.5$ and $1.645(\sigma / \sqrt{n_2}) \leq 1.0$. Algebraically, we find $\sqrt{n_1} \geq 1.645\sigma/1.5$, and $\sqrt{n_2} \geq 1.645\sigma/1.0$, and from the known sample size we have $1.645\sigma \leq 3.0\sqrt{100}$. Thus $\sqrt{n_1} \geq 3.0\sqrt{100}/1.5$, so $n_1 \geq 400$ and $\sqrt{n_2} \geq 3.0\sqrt{100}/1.0$, so $n_2 \geq 900$.

There are two points worth noting. First, what would have been the result if, for example, 95% had been used instead of 90% for the confidence level?

Answer: Each 1.645 would have been replaced by 1.96, and instead of dividing 1.645 by 1.645 we would divide 1.96 by 1.96, but the resulting answers of 400 and 900 would not have changed.

Second, how much must the sample size be increased in order to cut the interval estimate in half? To a third?

Answer: To cut the interval estimate in half (from ±3.0 to ±1.5), we would have to increase the sample size fourfold (from 100 to 400). To cut the interval estimate to a third (from ±3.0 to ±1.0), we would have to increase the sample size ninefold (from 100 to 900).

More generally, if we want to divide the interval estimate by d without effecting the confidence level, we must increase the sample size by a multiple of d^2.

Key 31 Hypothesis test of the mean

OVERVIEW *Closely related to the problem of estimating a population mean is the problem of testing a hypothesis about a population mean. For example, a consumer protection agency might determine an interval estimate for the mean nicotine content of a particular brand of cigarettes, or, alternatively, it might test a manufacturer's claim about the mean nicotine content of his cigarettes. An agricultural researcher could find an interval estimate for the mean productivity gain caused by a specific fertilizer, or, alternatively, she might test the developer's claimed mean productivity gain. A social scientist might ascertain an interval estimate for the mean income level of migrant farmers, or, alternatively, he might test a farm bureau's claim about the mean income level.*

The general testing procedure is to choose a specific hypothesis to be tested, called the **null hypothesis**, pick an appropriate sample, and use measurements from the sample to determine the likelihood of the null hypothesis. Conclusions are never stated with absolute certainty, but rather with associated significance levels. There are two types of possible errors that we consider: the error of mistakenly rejecting a true null hypothesis and the error of mistakenly failing to reject a false null hypothesis.

KEY EXAMPLE

A manufacturer claims that the mean lifetime of an electronic component of his product is 1500 hours. A researcher believes that the true figure is lower and will test the 1500-hour claim by measuring the lifetime of each element in a sampling of components. The researcher decides that, if the sample average is less than 1450 hours, she will reject the manufacturer's claim. Alternatively, if the sample average is over 1450 hours, she will conclude that she does not have sufficient evidence to reject the 1500-hour claim.

The claim to be tested, the *null hypothesis*, labeled H_0, is usually what we want to disprove, and is stated in terms of a specific value for a population parameter. In this case

$$H_0: \mu = 1500$$

The *alternative hypothesis*, denoted as H_a, is usually what we want to establish, and is stated in terms of an inequality such as <, >, or ≠. In this case

$$H_a: \mu < 1500$$

If the alternative hypothesis involves < or >, then there is one *critical value c* that separates the null hypothesis *rejection region* from the *fail to reject* region. In this case

$$c = 1450 \quad \text{and} \quad \frac{\text{rejection region} \mid \text{fail to reject region}}{1450}$$

If the 1500-hour claim is true but the sample mean happens to be less than 1450, then the researcher will *mistakenly reject* the null hypothesis. This is called a **Type I error**, and the probability of committing such an error is called the **α-risk.** If the 1500-hour claim is false but the sample mean happens to be over 1450, then the researcher will *mistakenly fail to reject* the null hypothesis. This is called a **Type II error**, and the probability of committing such an error is called the **β-risk.** (The Greek letters used to designate Type I and Type II errors are lowercase alpha and beta, respectively.)

Null Hypothesis ($\mu = 1500$)

		True	False
Sample Mean	Less than 1450	Type I error *mistakenly reject true claim*	Correct decision *reject false claim*
	More than 1450	Correct decision *do not reject true claim*	Type II error *mistakenly fail to reject false claim*

Several questions immediately come to mind. How should the critical value c be chosen? How can the probabilities of Type I and Type II errors be calculated? What effect does sample size have on the whole procedure?

Key 32 Critical values and α-risks

OVERVIEW *For a given critical value, we can sketch the distribution of sample means around the claimed population mean, and then note the region corresponding to the probability of a Type I error or α-risk. Alternatively, starting with an acceptable α-risk, we can calculate the corresponding critical value.*

KEY EXAMPLE

A patient claims that he is taking in only 2000 calories per day, but a dietician suspects that the actual figure is larger. The dietician plans to check food intake on 30 days and will reject the patient's claim if the 30-day mean is more than 2100 calories. If the standard deviation in calories per day is 350, what is the α- risk, that is, what is the probability of a Type I error?

Answer: We have:

$$H_0: \mu = 2000, \quad H_a: \mu > 2000$$

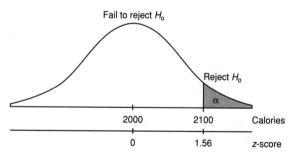

The standard deviation of sample means is $\sigma_{\bar{x}} = \sigma / \sqrt{n} = 350 / \sqrt{30} = 63.9$. The z-score of 2100 is $(2100 - 2000)/63.9 = 1.56$, and from Table A we obtain $.5000 - .4406 = .0594$. Thus, if the 2000-calorie claim is correct, there is a .0594 probability that the dietician will still obtain a mean greater than 2100 and mistakenly reject the patient's claim.

KEY EXAMPLE

A government statistician claims that the mean monthly rainfall along the Liberian coast is 15.0 inches with a standard deviation of 12.0 inches.

A meteorologist plans to test this claim with measurements over 3.5 years (42 months). If she finds a sample mean more than 2.0 inches different from the claimed 15.0 inches, she will reject the government statistician's claim. What is the probability that the meteorologist will mistakenly reject a true claim?

Answer: We have:

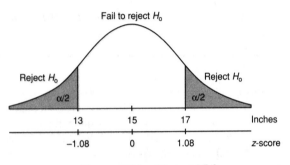

$$H_0: \mu = 15.0, \quad H_a: \mu \neq 15.0$$

This is an example of a two-sided (or two-tailed) test, and the α-risk is computed by adding the probabilities corresponding to two regions. The standard deviation of the sample means is $\sigma_{\bar{x}} = 12.0/\sqrt{42} = 1.85$. The z-score of 17 is $(17 - 15)/1.85 = 1.08$; similarly, the z-score of 13 is -1.08. From Table A we obtain $.5000 - .3599 = .1401$, for a total probability of Type I error of $.1401 + .1401 = .2802$.

The two examples given above begin with critical values and proceed to finding the corresponding α-risks. In most situations this reasoning is reversed; that is, one chooses acceptable α-risks and then calculates the corresponding critical values. The choice of α-risk is called the *significance level* of the test.

KEY EXAMPLE

A manufacturer claims that a particular automobile model will get 50 mpg on the highway. The staff of a consumer-oriented magazine believe that this claim is high and plan a test with a sample of 30 cars. Assuming the standard deviation between cars is 2.3 mpg, determine the critical value for a test at the 5% significance level.

Answer: We have:

$$H_0: \mu = 50, \quad H_a: \mu < 50, \quad \alpha = .05$$

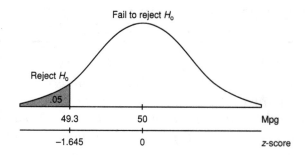

The standard deviation of sample means is $\sigma_{\bar{x}} = 2.3/\sqrt{30} = 0.42$. From Table A, the critical z-score corresponding to $\alpha = .05$ is -1.645, which translates to a raw score of $50 - 1.645(0.42) = 49.3$. Thus, if the sample mean is less than 49.3 mpg, the manufacturer's claim should be rejected.

KEY EXAMPLE

A student handbook claims that college students study an average of 30 hours per week. A guidance counselor plans to test this claim at the 1% significance level. What are the critical values if the standard deviation in student studying hours per week is 8 hours? Assume a sample of 100 students.

Answer: We have:

$$H_0: = \mu = 30, \quad H_a: \mu \neq 30, \quad \alpha = .01, \quad \text{and} \quad \sigma_{\bar{x}} = 8/\sqrt{100} = 0.8$$

This is a two sided (two tailed) test, and so for a total α-risk of .01 we put a probability of $\frac{1}{2}(.01) = .005$ on each side. A probability of .005 gives critical z-scores of ± 2.58. The corresponding critical hour values are $30 \pm 2.58 (0.8) = 30 \pm 2.06$. Thus the handbook claim of 30 hours should be rejected if the sample mean is less than 27.94 or greater than 32.06 hours.

Key 33   Drawing conclusions; *p*-values

OVERVIEW   *We attempt to show that the null hypothesis is incorrect by showing that it is improbable. Our testing procedure involves picking a sample and comparing the sample statistic (such as the mean $\bar{x}$) to the claimed population parameter (such as μ). Critical values are chosen to gauge the significance of a sample statistic. If the sample statistic is far enough away from the claimed population parameter, we say that there is sufficient evidence to reject the null hypothesis.*

KEY EXAMPLE

An automotive company executive claims that a mean number of 48.3 cars per dealership is being sold each month. A major stockholder believes that this claim is high and runs a test by sampling 30 dealerships. What conclusion is reached if the sample mean is 45.4 cars with a standard deviation of 15.4? Assume a 10% significance level.

Answer: We have:

$$H_0: \mu = 48.3, \quad H_a: \mu < 48.3, \quad \alpha = .10$$

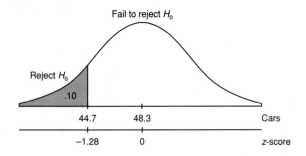

Here $\sigma_{\bar{x}} = \sigma / \sqrt{n} \approx s / \sqrt{n} = 15.4 / \sqrt{30} = 2.81$. With $\alpha = .10$, the critical z-score is -1.28. The critical number of cars is $48.3 - 1.28(2.81) = 44.7$. Since $45.4 > 44.7$, the stockholder should not reject the executive's claim.

If α had been chosen differently, would the executive's claim have been rejected?

Answer: We calculate that the sample statistic, 45.4, has a z-score of $(45.4 - 48.3)/2.81 = -1.03$. Using Table A, we note that if α was $.5000 - .3485 = .1515$ (or larger), the stockholder would have sufficient evidence to reject the 48.3-car claim.

The **p-value** of a test is the smallest value of α for which the null hypothesis would be rejected. Or equivalently, if the null hypothesis is assumed to be true, the p-value of a sample statistic is the probability of obtaining a result as extreme as the one obtained. Note that the smaller the p-value, the more significant is the difference between the null hypothesis and the sample results. In the above example, the p-value is .1515. Therefore, although the null hypothesis could not be rejected at the 10% (or even the 15%) significance level, it could be rejected at, for example, the 16% significance level.

KEY EXAMPLE

A manufacturer claims that a new brand of air-conditioning unit uses only 6.5 kilowatts of electricity per day. A consumer agency believes the true figure is higher and runs a test on a sample of size 50. If the sample mean is 7.0 kilowatts with a standard deviation of 1.4, should the manufacturer's claim be rejected at a significance level of 5%? Of 1%? What is the p-value of this test result?

Answer: We have:

$$H_0: \mu = 6.5, \quad H_a: \mu > 6.5, \quad \alpha = .05$$

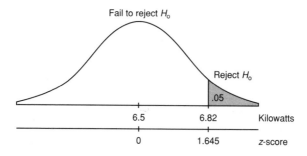

Here $\sigma_{\bar{x}} = \sigma/\sqrt{n} \approx s/\sqrt{n} = 1.4/\sqrt{50} = 0.198$. With $\alpha = .05$, the critical z-score is 1.645. There are actually two ways to proceed. We could convert 1.645 to a critical number of kilowatts and check whether 7.0 is greater than this critical number, or we could convert 7.0 to a z-score and compare this to the critical z-score 1.645. By the first method, 1.645 converts to a raw score of $6.5 + 1.645(0.198) = 6.83$;

7.0 > 6.83, so reject H_0. By the second method, 7.0 converts to a z-score of $(7.0 - 6.5)/0.198 = 2.5$; $2.53 > 1.645$, so reject H_0. Thus, at the 5% significance level, the consumer agency should reject the manufacturer's claim.

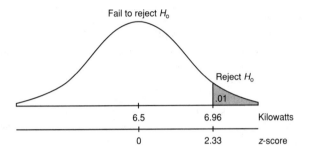

How about $\alpha = .01$? The critical z-score is now 2.33 with a corresponding critical number of kilowatts: $6.5 + 2.33(0.198) = 6.96$. Since $7.0 > 6.96$, the consumer should still reject the manufacturer's claim at the 1% significance level. Note that we could have compared 2.53, the z-score of 7.0, with 2.33 to reach the same conclusion.

We have shown that the difference between the sample statistic ($\bar{x} = 7.0$) and the claimed population parameter ($\mu = 6.5$) is significant at both the .05 and .01 levels. Calculation of the p-value gives a more complete picture of the significance of the observed difference.

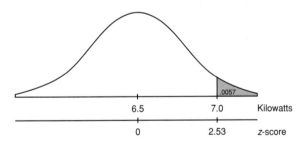

The z-score of 7.0 is 2.53; using Table A, we find the p-value to be $.5000 - .4943 = .0057$. Note that with this calculation we could have quickly concluded that at the .05 and .01 significance levels there is sufficient evidence to reject H_0. We also see that there would have been sufficient evidence at the .006 level, but not at the .005 level.

KEY EXAMPLE

A local chamber of commerce claims that the mean family income level is $12,250. An economist runs a hypothesis test, using a sample of 135 families, and finds a mean of $11,500 with a standard deviation of $3180. Should the $12,250 claim be rejected at a 5% level of significance?

Answer: Note that this is a two-sided test; that is, the economist suspects that the claim may be incorrect, but does not know whether it is probably too high or too low. We have:

$$H_0: \mu = 12{,}250, \quad H_a: \mu \neq 12{,}250, \quad \alpha = .05$$

Here $\sigma_{\bar{x}} = \sigma / \sqrt{n} \approx s / \sqrt{n} = 3180 / \sqrt{135} = 273.7$. For a total $\alpha = .05$, we place a probability of $.5(.05) = .025$ on each side, and note the critical z-scores of ± 1.96. These translate into critical incomes of $12{,}250 \pm 1.96(273.7)$ or $11{,}713.45$ and $12{,}786.45$. Since $11{,}500 < 11{,}713.45$, the chamber of commerce claim is rejected.

Note that, alternatively, we could have changed $11{,}500$ into a z-score: $(11{,}500 - 12{,}250)/273.7 = -2.74$ and observed that -2.74 is less than the critical z-score, -1.96. We could have also reasoned in terms of the p-value: For the z-score, -2.74, Table A gives a probability of $.5000 - .4969 = .0031$. Since this is a two-tailed test, where disagreement with the null hypothesis can be in either of two directions, the p-value is twice the $.0031$ probability, or $.0062$. Since $.0062 < .05$, there is sufficient evidence to reject H_0.

Key 34 Type II errors

OVERVIEW *Why not always choose the α-risk to be extremely small, such as .001 or .0001, and so virtually eliminate the possibility of mistakenly rejecting a correct null hypothesis? The problem is that doing this would simultaneously increase the chance of never rejecting the null hypothesis, even if it were false. Thus we are led to a discussion of the Type II error, that is, a mistaken failure to reject a false null hypothesis. The probability of a Type II error is called the β-risk, and there is a different β-value for each possible correct value for the population parameter.*

KEY EXAMPLE

City planners are trying to decide among various parking-plan options ranging from more on-street spaces to multilevel facilities to spread-out small lots. Before making a decision, they wish to test the downtown merchants' claim that shoppers park for an average of only 47 minutes in the downtown area. The planners have decided to tabulate parking durations for 225 shoppers and to reject the merchants' claim if the sample mean is over 50 minutes. What is the probability of a Type II error if the true value is 48? If the true value is 51? Assume that the standard deviation in parking durations is 27 minutes.

Answer: We have:

$$H_0: \mu = 47, \quad H_a: \mu > 47$$

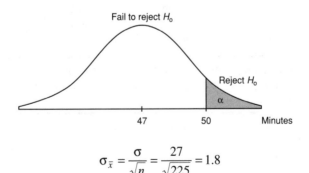

$$\sigma_{\bar{x}} = \frac{\sigma}{\sqrt{n}} = \frac{27}{\sqrt{225}} = 1.8$$

If the true mean parking duration is 48, then the normal curve should be centered at 48. In this case, the z-score of 50 is $(50 - 48)/1.8 = 1.11$. Using Table A, we calculate the β-risk (probability of failure to reject H_0) to be $.5000 + .3665 = .8665$.

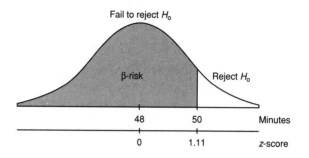

If the true mean parking duration is 51, then the normal curve should be centered at 51. The critical value is still 50, with a z-score now of $(50 - 51)/1.8 = -.56$. We use Table A to find the β-risk to be $.5000 - .2123 = .2877$.

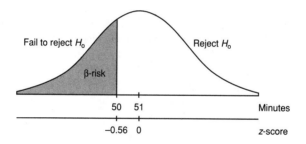

Note that the further the true value is (in the suspected direction) from the claimed mean, the smaller the probability is of failing to reject the false claim.

In many situations we start with a significance level, use this to calculate the critical value, and then find particular β-risks.

KEY EXAMPLE

A geologist claims that a particular rock formation will yield a mean of 24 pounds of a chemical per ton of excavation. A company, fearful that the true amount will be less, plans to run a test on a sample of 50 tons.

If the standard deviation between tons is 5.8 pounds, what is the critical value? Assume a 1% significance level. What is the probability of a Type II error if the true mean is 22? Is 20?

Answer: We have:

$$H_0: \mu = 24, \quad H_a: \mu < 24, \quad \alpha = .01$$

$$\sigma_{\bar{x}} = \frac{\sigma}{\sqrt{n}} = \frac{5.8}{\sqrt{50}} = 0.82$$

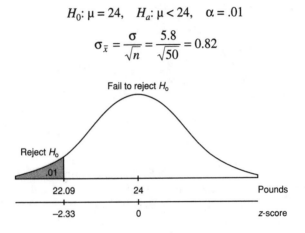

With $\alpha = .01$ the critical z-score is -2.33, and the critical poundage value is $24 - 2.33(0.82) = 22.09$. Thus, the decision rule is to reject H_0 if the sample mean is less than 22.09, and fail to reject if it is more than 22.09.

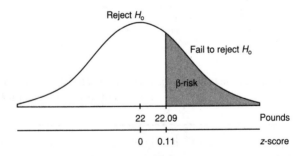

If the true mean is 22 pounds per ton of rock, then the z-score of 22.09 is $(22.09 - 22)/0.82 = 0.11$, and so the β-risk is $.5000 - .0438 = .4562$.

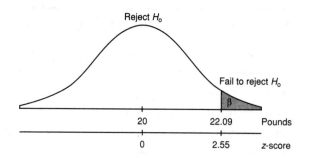

If the true mean is 20 pounds per ton of rock, then the z-score of 22.09 is $(22.09 - 20)/0.82 = 2.55$, and so the β-risk is $.5000 - .4946 = .0054$.

KEY EXAMPLE

A factory manager claims that the plant's smokestacks spew forth only 350 pounds of pollution per day. A government investigator suspects that the true value is higher and plans a hypothesis test with a critical value of 375 pounds. Suppose the standard deviation in daily pollution poundage is 150. What is the α-risk for a sample size of 100 days? 200 days? What are the associated β-risks if the true mean is 385 pounds?

Answer: We have: H_0: $\mu = 350$, H_a: $\mu > 350$, and $c = 375$

With $n = 100$, $\sigma_{\bar{x}} = 150/\sqrt{100} = 15.0$, the z-score of 375 is $(375 - 350)/15.0 = 1.67$, and the corresponding α-risk from Table A is $.5000 - .4525 = .0475$. With $n = 200$, $\sigma_{\bar{x}} = 150/\sqrt{200} = 10.6$, the z-score of 375 is $(375 - 350)/10.6 = 2.36$, and the corresponding α-risk is $.5000 - .4909 = .0091$.

With $n = 100$ and a mean of 358, the z-score of 375 is $(375 - 385)/15.0 = -0.67$ with a corresponding β-risk of $.5000 - .2486 = .2514$. With $n = 200$ and a mean of 385, the z-score of 375 is $(375 - 385)/10.6 = -0.94$ with a corresponding β-risk of $.5000 - .3264 = .1736$.

We note that in this particular example *increasing* the sample size (from 100 to 200) resulted in *decreases* in both the α-risk (from .0475 to .0091) and the β-risk (from .2514 to .1736). More generally, if α is held fixed and the sample size is increased, then the β-risk will decrease. Furthermore, by increasing the sample size and adjusting the critical value, it is always possible to decrease both the α-risk and the β-risk.

Key 35   Confidence interval estimate for the difference of two means

OVERVIEW   *Many real-life applications of statistics involve comparisons of two populations. For example, is the average weight of laboratory rabbits receiving a special diet greater than that of rabbits on a standard diet, which of two accounting firms gives a higher mean starting salary, is the life expectancy of a coal miner less than that of a school teacher? To compare the means of two populations, we compare the means of two samples, one from each population.*

In many situations it is clear that the mean of one population is higher than that of another, and we would like to estimate the difference. Using samples, we cannot find this difference *exactly*, but we will be able to say with a certain confidence that the difference lies in a certain *interval*. We follow the same procedure as set forth for a single mean, this time using $\mu_1 - \mu_2$, $\sqrt{\sigma_1^2/n_1 + \sigma_2^2/n_2}$, and $\bar{x}_1 - \bar{x}_2$ in place of μ, $\sigma/\sqrt{n}$, and $\bar{x}$, respectively. [Remark: the formula for standard deviation is derived from the fact that the variance of a set of differences is equal to the sum of the variances of the individual sets.]

KEY EXAMPLE

A 30-month study is conducted to determine the difference in the rates of accidents per month between two departments in an assembly plant. Suppose the first department averaged 12.3 accidents per month with a standard deviation of 3.5, while the second averaged 7.6 accidents with a standard deviation of 3.4. Determine a 95% confidence interval estimate for the difference in accidents per month.

Answer:

$$n_1 = 30 \qquad n_2 = 30$$
$$\bar{x}_1 = 12.3 \qquad \bar{x}_2 = 7.6$$
$$s_1 = 3.5 \qquad s_2 = 3.4$$

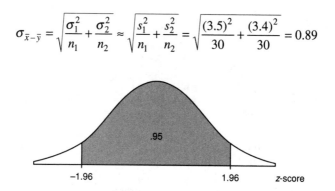

$$\sigma_{\bar{x}-\bar{y}} = \sqrt{\frac{\sigma_1^2}{n_1} + \frac{\sigma_2^2}{n_2}} \approx \sqrt{\frac{s_1^2}{n_1} + \frac{s_2^2}{n_2}} = \sqrt{\frac{(3.5)^2}{30} + \frac{(3.4)^2}{30}} = 0.89$$

The observed difference is $12.3 - 7.6 = 4.7$, and the critical z-scores are ± 1.96. Thus the confidence interval estimate is $4.7 \pm 1.96(0.89) = 4.7 \pm 1.74$. We are 95% confident that the first department has between 2.96 and 6.44 more accidents per month than does the second department.

KEY EXAMPLE

A survey is run to determine the difference in the cost of groceries in suburban stores versus inner-city stores. A preselected group of items is purchased in a sample of 45 suburban and 35 inner-city stores, and the following data are obtained:

Suburban stores	Inner-city stores
$n_1 = 45$	$n_2 = 35$
$\bar{x}_1 = \$36.52$	$\bar{x}_2 = \$39.40$
$s_1 = \$1.10$	$s_2 = \$1.23$

Find a 90% confidence interval estimate for the difference.

Answer:

$$\sigma_{\bar{x}-\bar{y}} = \sqrt{\frac{(1.10)^2}{45} + \frac{(1.23)^2}{35}} = 0.265$$

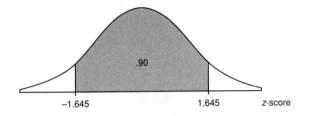

The observed difference is $36.52 - 39.40 = -2.88$, the critical z-scores are ± 1.645, and the confidence interval estimate is $-2.88 \pm 1.645(0.265) = -2.88 \pm 0.44$. Thus we are 90% certain that the selected group of items costs between \$2.44 and \$3.32 *less* in suburban stores than in inner-city stores.

A common sample size can be chosen by a method similar to that used for the one-sample case.

KEY EXAMPLE

A hardware store owner wishes to determine the difference in drying times between two brands of paints. Suppose the standard deviation between cans in each population is 2.5 minutes. How large a sample (same number) of each must be used if the owner wishes to be 98% sure of knowing the difference to within 1 minute?

Answer:

$$\sigma_{\bar{x}-\bar{y}} = \sqrt{\frac{(2.5)^2}{n} + \frac{(2.5)^2}{n}} = \frac{3.536}{\sqrt{n}}$$

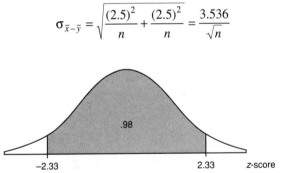

With a critical z-score of 2.33, we have $2.33\left(3.536/\sqrt{n}\right) \leq 1$, so $\sqrt{n} \geq 8.24$ and $n \geq 67.9$. Thus the owner should test samples of 68 paint patches from each brand.

Key 36 Hypothesis test for the difference of two means

OVERVIEW   *The fact that a sample mean from one population is greater than a sample mean from a second population does not automatically lead to a similar conclusion about the means of the populations themselves. Two points need to be stressed. First, means of samples from the same population vary from each other. Second, what we are really comparing are confidence interval estimates, not just single points.*

In this situation the null hypothesis is usually that the means of the populations are the same, or, equivalently, that their difference is 0:

$$H_0: \mu_1 - \mu_2 = 0$$

The alternative hypothesis is then:

$$H_a: \mu_1 - \mu_2 < 0, \quad H_a: \mu_1 - \mu_2 > 0, \quad \text{or} \quad H_a: \mu_1 - \mu_2 \neq 0$$

The first two possibilities lead to one-sided (one-tailed) tests, while the third possibility leads to two-sided (two-tailed) tests.

KEY EXAMPLE

A salesman believes that his company's computer has more average downtime per week than does a similar computer manufactured by a competitor. Before bringing his concern to his director, the salesman gathers data and runs a hypothesis test. In a sample of 40 week-long periods in different firms using his company's product, the average downtime was 125 minutes with a standard deviation of 37 minutes. However, 35 week-long periods involving the competitor's computer yield an average downtime of only 115 minutes with a standard deviation of 43 minutes. What conclusion should be drawn, assuming a 10% significance level?

Answer:

$$H_0: \mu_1 - \mu_2 = 0, \quad H_a: \mu_1 - \mu_2 > 0, \quad \alpha = .10$$

$$\sigma_{\bar{x}-\bar{y}} = \sqrt{\frac{(37)^2}{40} + \frac{(43)^2}{35}} = 9.33$$

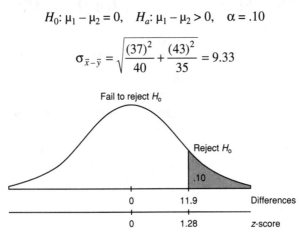

With $\alpha = .10$ the critical z-score is 1.28, so the critical difference is $0 + 1.28(9.33) = 11.9$ minutes. Since the observed difference is $125 - 115 = 10$, and $10 < 11.9$, the salesman does *not* have sufficient evidence to say that his company's computer has more downtime.

The strength of the observed difference in downtimes can be measured by finding the *p*-value. The z-score of the difference in sample means is $10/9.33 = 1.07$. Using Table A, we find the *p*-value to be $.5000 - .3577 = .1423$. Thus, while the observed difference is not significant at the 10% significance level, it would be significant at, for example, the 15% level.

KEY EXAMPLE

A store manager wishes to determine whether there is a significant difference between two trucking firms with regard to the handling of egg cartons. In a sample of 200 cartons on one firm's truck, there was an average of 0.7 broken egg per carton with a standard deviation of 0.31, while a sample of 300 cartons on the second firm's truck showed an average of 0.775 broken egg per carton with a standard deviation of 0.42. Is the difference between the averages significant at a significance level of 5%? At a level of 1%?

Answer:

$$H_0: \mu_1 - \mu_2 = 0, \quad H_a: \mu_1 - \mu_2 \neq 0, \quad \alpha = .05$$

$$\sigma_{\bar{x}-\bar{y}} = \sqrt{\frac{(0.31)^2}{200} + \frac{(0.42)^2}{300}} = 0.0327$$

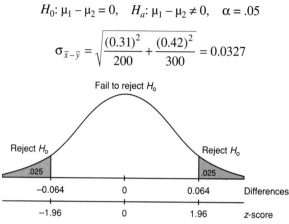

Note that this is a two-tailed test, so for a total $\alpha = .05$ we place a probability of $.5(.05) = .025$ on each side. The critical z-scores are ± 1.96, which translate into critical differences of $0 \pm 1.96(0.0327) = \pm 0.064$. Since the observed difference is $0.7 - 0.775 = -0.075$ and $-0.075 < -.064$, the difference in the average numbers of broken eggs per carton is significant at the 5% significance level.

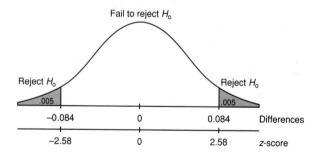

With $\alpha = .01$, the critical z-scores are ± 2.58, and so the critical differences are $0 \pm 2.58(0.0327) = \pm 0.084$. Since -0.075 is between -0.084 and 0.084, the observed difference is *not* significant at the 1% significance level.

A more specific measure of the strength of the observed difference in broken eggs per carton is found by calculating the p-value. The observed difference of -0.075 has a z-score of $-0.075/0.0327 = -2.29$. Using Table A, we find that this gives a probability of $.5000 - .4890 = .0110$. The test is two-sided, so the p-value is $2(.0110) = .0220$.

Key 37   Theme exercises with answers

OVERVIEW   *Sample questions of the type that might appear on homework assignments and tests are presented with answers.*

- In a certain plant, batteries are being produced with life expectancies that have a variance of 5.76 months^2. Suppose the mean life expectancy in a sample of 64 batteries is 12.35 months. Find a 90% confidence interval estimate of life expectancy for all batteries produced in this plant.

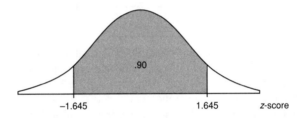

Answer: The standard deviation of the population is $\sigma = \sqrt{5.76} = 2.4$, and the standard deviation of the sample means is $\sigma_{\bar{x}} = \sigma/\sqrt{n} = 2.4/\sqrt{64} = 0.3$. The 90% confidence interval estimate for the population mean is $12.35 \pm 1.645(0.3) = 12.35 \pm 0.4935$. Thus we are 90% certain that the mean life expectancy of the batteries is between 11.8565 and 12.8435 months.

What would be the 90% confidence interval estimate if the sample mean of 12.35 had come from a sample of 100 batteries?

Answer: The standard deviation of sample means would then have been $\sigma_{\bar{x}} = \sigma/\sqrt{n} = 2.4/\sqrt{100} = 0.24$, and the 90% confidence interval estimate would be $12.35 \pm 1.645(0.24) = 12.35 \pm 0.3948$.

Note that, when the sample size increased (from 64 to 100), the same sample mean resulted in a narrower, more specific interval (± 0.3948 instead of ± 0.4935).

- A government investigator plans to test for the mean quantity of a particular pollutant that a manufacturer is dumping per day into a river. She needs an estimate that is within 50 grams at the 95% confidence level. If previous measurements indicate that the variance is approximately 21,800 grams^2, how many days should she include in the sample?

Answer:

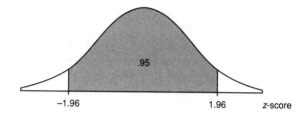

.95

−1.96   1.96   *z*-score

We have $\sigma^2 = 21,800$, so $\sigma = \sqrt{21,800} = 147.65$. Then $\sigma_{\bar{x}} = \sigma/\sqrt{n} = 147.65/\sqrt{n}$, and $1.96\sigma_{\bar{x}} \leq 50$. Thus $1.96\left(147.65/\sqrt{n}\right) \leq 50$, $\sqrt{n} \geq 1.96(147.65)/50 = 5.788$, and $n \geq 33.5$. Therefore, the investigator should sample 34 days' output.

• A coffee-dispensing machine is supposed to drop 8 ounces of liquid into each paper cup, but a consumer believes that the actual amount is less. As a test he plans to obtain a sample of 36 cups of the dispensed liquid, and, if the mean content is less than 7.75 ounces, to reject the 8-ounce claim. If the machine operates with a standard deviation of 0.9 ounce, what is the α-risk?

Answer: We have:

$$H_0: \mu = 8, \quad H_a: \mu < 8$$

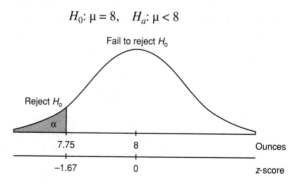

Fail to reject H_o

Reject H_o

α

7.75   8   Ounces

−1.67   0   *z*-score

The standard deviation of sample means is $\sigma_{\bar{x}} = \sigma/\sqrt{n} = 0.9/\sqrt{36} = 0.15$. The *z*-score of 7.5 is (7.75 − 8)/0.15 = −1.67. Using Table A, we obtain α = .5000 − .4525 = .0475. Thus, if the 8-ounce claim is correct, there is a .0475 probability that the consumer will still obtain a sample mean less than 7.75 and mistakenly reject the claim.

- A pharmaceutical company claims that a medication will produce a desired effect for a mean of 58.4 minutes. A government researcher runs a hypothesis test of 250 patients and tabulates the following data with reference to the durations of the effect in minutes: $\Sigma x = 14{,}875$ and $\Sigma(x - \bar{x})^2 = 17{,}155$. Should the company's claim be rejected at a significance level of 10%? Of 2%?

Answer:

$$\bar{x} = \frac{\Sigma x}{n} = \frac{14875}{250} = 59.5, \qquad s = \sqrt{\frac{\Sigma(x - \bar{x})^2}{n-1}} = \sqrt{\frac{17155}{249}} = 8.3$$

and

$$\sigma_{\bar{x}} \approx \frac{s}{\sqrt{n}} = \frac{8.3}{\sqrt{250}} = 0.525$$

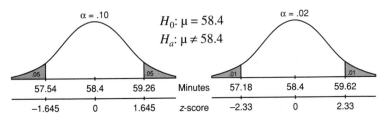

For $\alpha = .10$, the critical values in minutes are $58.4 \pm 1.645(0.525)$, or 57.54 and 59.26. For $\alpha = .02$, the critical values in minutes are $58.4 \pm 2.33(0.525)$, or 57.18 and 59.62. Now $59.5 > 59.26$, but $59.5 < 59.62$, so the researcher would reject the company's claim at the 10% significance level, but not at the 2% significance level. In other words, if she is willing to be wrong 10 times in 100, she would reject the company's claim; but if she is willing to accept only 2 errors in 100 such decisions, she would conclude that the evidence is not strong enough to reject the claim.

Again, the *p*-value can be used to specifically measure the strength of evidence for rejection of H_0. In this example, the z-score of 59.5 is $(59.5 - 58.4)/0.525 = 2.10$. Using Table A, we find the corresponding probability to be $.5000 - .4821 = .0179$. Doubling because the test is two-sided results in a *p*-value of $2(.0179) = .0358$.

- A medical research team claims that high vitamin C intake increases endurance. In particular, 1000 mg of vitamin C per day for a month should add an average of 4.3 minutes to the possible length of maximum physical effort. Army training officers believe the claim is exaggerated and plan a test on 400 soldiers. If the standard deviation of added minutes is 3.2, find the critical values for a significance level of 5% and then of 1%. In each case find also the probability of a Type II error if the true mean increase is only 4.2 minutes.

Answer: We have:

$$H_0: \mu = 4.3, \quad H_a: \mu < 4.3$$

$$\sigma_{\bar{x}} = \frac{3.2}{\sqrt{400}} = 0.16$$

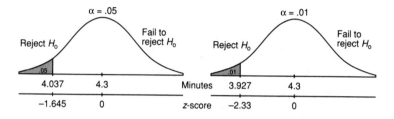

Corresponding to a z-score of -1.645 is $4.3 - 1.645(0.16) = 4.037$; corresponding to a z-score of -2.33 is $4.3 - 2.33(0.16) = 3.927$.

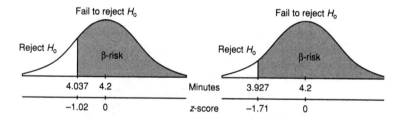

With a mean of 4.2, the z-score of 4.038 is $(4.038 - 4.2)/0.16 = -1.02$, and the β-risk is $.5000 + .3461 = .8461$. With a mean of 4.2, the z-score of 3.927 is $(3.927 - 4.2)/0.16 = -1.71$, and the β-risk is $.5000 + .4594 = .9564$.

We note that *decreasing* the α-risk (from .05 to .01) led to an *increase* in the β-risk (from .8438 to .9564). This is always the case if the sample size is held constant.

- A trucking firm conducts a test to compare the life expectancies of two brands of tires. Shipments of 1000 and 1500 tires, respectively, are received and marked, and the truck mileages are noted when the tires are replaced. The resulting raw data are as follows:

Brand F	Brand G
$n_1 = 1000$	$n_2 = 1500$
$\sum x = 22,350,000$	$\sum x = 36,187,500$
$\sum(x - \bar{x})^2 = 9,600,000,000$	$\sum(x - \bar{x})^2 = 15,800,000,000$

Determine a 99% confidence interval estimate for the difference in life expectancies.

Answer: Calculations of the means and standard deviations yield:

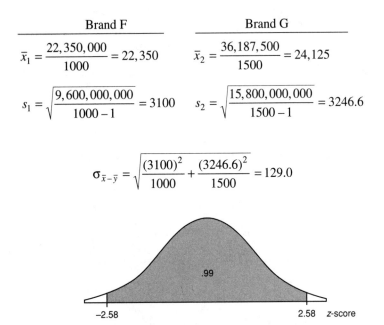

Brand F	Brand G
$\bar{x}_1 = \dfrac{22,350,000}{1000} = 22,350$	$\bar{x}_2 = \dfrac{36,187,500}{1500} = 24,125$
$s_1 = \sqrt{\dfrac{9,600,000,000}{1000 - 1}} = 3100$	$s_2 = \sqrt{\dfrac{15,800,000,000}{1500 - 1}} = 3246.6$

$$\sigma_{\bar{x}-\bar{y}} = \sqrt{\frac{(3100)^2}{1000} + \frac{(3246.6)^2}{1500}} = 129.0$$

.99

−2.58   2.58   z-score

The observed difference is $22,350 - 24,125 = -1775$, and the critical z–scores are ± 2.58. Since $-1775 \pm 2.58(129) = -1775 \pm 333$, the trucking firm can be 99% sure that brand F tires average between 1442 and 2108 miles less in life expectancy than brand G tires.

- A realtor believes that the average rent in one location is greater than the published figure of $125 over the average rent in a second location. She runs a hypothesis test on 30 rental units in each location and obtains averages of $480 and $330, and standard deviations of $95 and $80, respectively, for the two locations. What should the realtor conclude at the 5% significance level?

Answer: Note that in this problem the null hypothesis is not that the difference is 0. We have:

$$H_0: \mu_1 - \mu_2 = 125, \quad H_a: \mu_1 - \mu_2 > 125, \quad \alpha = .05$$

$$\sigma_{\bar{x}-\bar{y}} = \sqrt{\frac{(95)^2}{30} + \frac{(80)^2}{30}} = 22.68$$

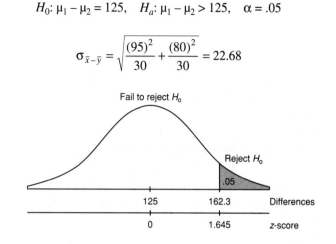

With $\alpha = .05$ the critical z-score is 1.645, and so the critical difference is $125 + 1.645(22.68) = 162.3$. The observed difference is $480 - 330 = 150$. Since $150 < 162.3$, the realtor does *not* have sufficient evidence to claim that the rental difference is more than 125.

There was some observed difference, and its strength can be measured by finding the p-value. The z-score of the observed difference of 150 is $(150 - 125)/22.68 = 1.10$. Using Table A, we find the p-value to be $.5000 - .3643 = .1357$.

Theme 6   SMALL SAMPLES

*N*umerous man-hours would be saved if small samples could be used instead of larger ones. There are many situations, such as estimating the completion time for a certain model nuclear reactor or determining the damage sustained by luxury cars from front-end collisions, when it is prohibitively difficult or expensive to gather enough data for a large sample. In other situations, such as deciding whether to add or drop a TV series after only a few shows have been aired, there is simply not sufficient time to gather large amounts of data.

Confidence intervals estimating and hypothesis testing using small samples involve the *Student* t-*distribution,* introduced in Key 38. In this situation, unlike the case with large samples, certain assumptions are required concerning the populations under consideration. Furthermore, conclusions using small samples involve larger dispersions of sample means.

Key 38   Student *t*-distributions

OVERVIEW   *For working with small samples, the appropriate distribution is the* **Student t-distribution**, *developed in 1908 by W. S. Gossett, a British mathematician employed by the Guiness Breweries. This distribution is bell shaped and symmetrical, but is lower at the mean, higher at the tails, and so more spread out than the normal distribution.*

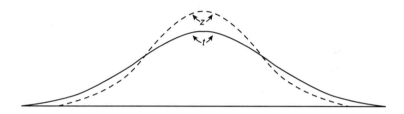

In the case of small samples, we can no longer say, for example, that 95% of the sample means are within $1.96\sigma_{\bar{x}}$ of the population mean. More than 5% of the sample means will lie outside these critical values. However, if the original population is normally distributed, then the variable

$$t = \frac{\bar{x} - \mu}{\sigma_{\bar{x}}}$$

has the resulting distribution pictured above.

Like the binomial distribution, the *t*-distribution is different for different values of *n*. In tables these distinct *t*-distributions are associated with the value of df (degrees of freedom), which in our present consideration is equal to the sample size minus 1. The smaller the df value, the larger is the dispersion in the distribution. The larger the df value, that is, the larger the sample size, the closer the distribution comes to the normal distribution.

Since there is a separate *t*-distribution for each degree of freedom, fairly complete tables would involve many pages; therefore, we give areas and t-values for only the more commonly used percentages or probabilities. The last row of Table B in the Appendix is the normal distribution because the normal distribution is a special case of the t-distribution taken when n is infinite. For practical purposes, the two distributions are very close for any $n \geq 30$.

Note that, whereas Table A gives areas under the normal curve from the mean to positive z values, Table B gives areas to the right of given positive t values. For example, suppose the sample size is 20 and so df $= 20 - 1 = 19$. Then a probability of .05 in the *tail* will correspond with a t value of 1.729, while .01 in the tail corresponds to $t = 2.539$.

Key 39 Confidence interval estimate
of the mean

OVERVIEW *Just as with large samples, the sample mean is used to state with a certain confidence that the population mean lies in a certain interval. However, having to use the* t-*distribution instead of the* z-*distribution leads to less precise conclusions.*

In the illustrations throughout Theme 6, there is the assumption that the original populations are normally distributed. Without this, the *t*-distribution could not be used.

KEY EXAMPLE

When ten cars of a new model were tested for gas mileage, the results showed a mean of 27.2 mpg with a standard deviation of 1.8 mpg. A 95% confidence interval estimate for the miles per gallon achieved by this model is obtained as follows:

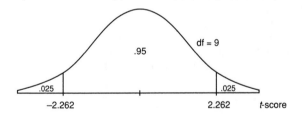

Because the sample size is below 30, we must use the *t*-distribution. The standard deviation of the sample means $\sigma_{\bar{x}} = 1.8/\sqrt{10} = 0.569$. With $10 - 1 = 9$ degrees of freedom, and 2.5% in each tail, the appropriate *t*-scores are ±2.262. Thus we can be 95% certain that the gas mileage of the new model is in the range $27.2 \pm 2.262(0.569) = 27.2 \pm 1.3$, or between 25.9 and 28.5 mpg.

KEY EXAMPLE

Twenty-five "18 ounce" jars of peanut butter are weighed, yielding the totals $\sum x = 448.5$ and $\sum(x-\bar{x})^2 = 0.41$. What is the 99% confidence interval estimate for the mean weight?

Answer: We first calculate the sample mean and standard deviation:

$$\bar{x} = \frac{\sum x}{n} = \frac{448.5}{25} = 17.94, \qquad s = \sqrt{\frac{\sum(x-\bar{x})^2}{n-1}} = \sqrt{\frac{0.41}{24}} = 0.13$$

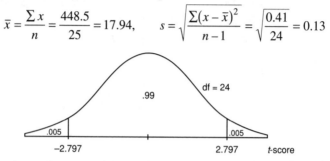

The standard deviation $\sigma_{\bar{x}}$ of the sample means is estimated to be $s/\sqrt{n} = 0.13/\sqrt{25} = 0.026$. With $25 - 1 = 24$ degrees of freedom, and 0.5% in each tail, the appropriate t-scores are ±2.797. Thus a 99% confidence interval estimate is given by $17.94 \pm 2.797(0.026) = 17.94 \pm 0.073$ ounces.

KEY EXAMPLE

A new process for producing synthetic gems yielded, in its first run, six stones weighing 0.43, 0.52, 0.46, 0.49, 0.60, and 0.56 carat, respectively. Find a 90% confidence interval estimate for the mean carat weight from this process.

Answer:

$$\bar{x} = \frac{\sum x}{n} = \frac{0.43 + 0.52 + 0.46 + 0.49 + 0.60 + 0.56}{6} = \frac{3.06}{6} = 0.51$$

$$
\begin{aligned}
s &= \sqrt{\frac{\sum(x-\bar{x})^2}{n-1}} \\
&= \sqrt{\frac{(0.08)^2 + (0.01)^2 + (0.05)^2 + (0.02)^2 + (0.09)^2 + (0.05)^2}{5}} \\
&= 0.0632
\end{aligned}
$$

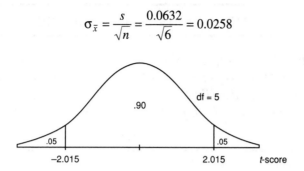

$$\sigma_{\bar{x}} = \frac{s}{\sqrt{n}} = \frac{0.0632}{\sqrt{6}} = 0.0258$$

With df = 6 − 1 = 5, and 5% in each tail, the t-scores are ±2.015. Thus we can be 90% sure that the new process will yield stones weighing 0.51 ± 2.015(0.0258) = 0.51 ± 0.052, or between 0.458 and 0.562 carat.

Key 40 Hypothesis test of the mean

OVERVIEW *Using a small sample to test a hypothesis about a population mean involves calculating the critical value from the* t-*distribution. Otherwise the procedure is the same as that worked out for large samples.*

KEY EXAMPLE

A cigarette industry spokesperson remarks that current levels of tar are no more than 5 milligrams per cigarette. A reporter does a quick check on 15 cigarettes representing a cross section of the market. What conclusion is reached if the sample mean is 5.63 milligrams of tar with a standard deviation of 1.61? Assume a 10% significance level.

Answer:

$$H_0: \mu = 5, \quad H_a: \mu > 5, \quad \alpha = .10$$

$$\sigma_{\bar{x}} = \frac{s}{\sqrt{n}} = \frac{1.61}{\sqrt{15}} = 0.42$$

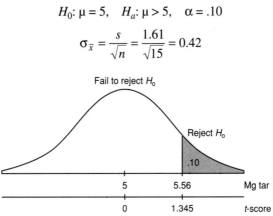

With df = 15 − 1 = 14, and α = .10, the critical *t*-score is 1.345. The critical number of milligrams of tar is 5 + 1.345(0.42) = 5.56. Since 5.63 > 5.56, the industry spokesperson's remarks should be rejected at the 10% significance level.

What is the conclusion at the 5% significance level?

Answer: In this case, the critical *t*-score is 1.761; the critical tar level is $5 + 1.761(0.42) = 5.74$; and since $5.63 < 5.74$, the remarks cannot be rejected at the 5% significance level.

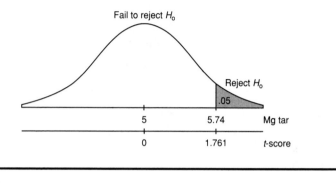

Fail to reject H_0

Reject H_0

.05

| 5 | 5.74 | Mg tar |
| 0 | 1.761 | *t*-score |

KEY EXAMPLE

A local chamber of commerce claims that the mean sale price for homes in the city is $90,000. A real estate salesperson notes eight recent sales of $75,000, $102,000, $80,000, $85,000, $79,000, $95,000, $98,000, and $62,000. Should the chamber of commerce's claim be rejected at a significance level of 5%? At a level of 10%?

Answer:
$$H_0: \mu = 90,000, \quad H_a: \mu \neq 90,000$$

$$\bar{x} = \frac{\sum x}{n} = \frac{676,000}{8} = 84,500$$

$$s = \sqrt{\frac{\sum(x - \bar{x})^2}{n-1}} = \sqrt{\frac{1,246,000,000}{7}} = 13,341.7$$

$$\sigma_{\bar{x}} = \frac{s}{\sqrt{n}} = \frac{13,341.7}{\sqrt{8}} = 4717$$

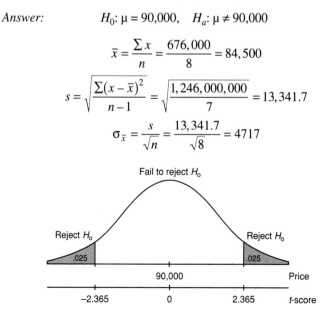

Fail to reject H_0

Reject H_0

Reject H_0

.025

.025

| | 90,000 | | Price |
| -2.365 | 0 | 2.365 | *t*-score |

For $\alpha = .05$, we place a probability of .025 on each side, and note that df = 8 − 1 = 7 gives critical t-scores of ±2.365. Critical sales prices are 90,000 ± 2.365(4717) = 90,000 ± 11,156 or \$78,844 and \$101,156. Since the observed mean of \$84,500 is in this range, there is not sufficient evidence to reject the claim at the 5% significance level.

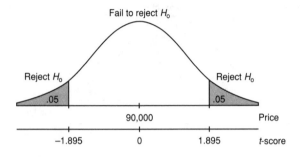

For $\alpha = .10$, we place .05 in each tail, and note critical t-scores of ±1.895. Critical sales prices are 90,000 ± 1.895(4717) = 90,000 ± 8939 or \$81,061 and \$98,939. The observed mean of \$84,500 is also in this range, so even at the 10% significance level the claim cannot be rejected.

Key 41 Differences in means

OVERVIEW *Small samples from two populations can be used to obtain a confidence interval estimate of the difference of the population means and to run a hypothesis test concerning the difference of the population means.*

Here, in addition to assuming both original populations are normally distributed, we must also assume that their variances are equal. Then for $\sigma_{\bar{x}-\bar{y}}$ (see Key 35) we get the complex-looking formula

$$\sigma_{\bar{x}-\bar{y}} = \sqrt{\frac{(n_1-1)s_1^2 + (n_2-1)s_2^2}{n_1+n_2-2}}\sqrt{\frac{1}{n_1}+\frac{1}{n_2}}$$

The number of degrees of freedom in such a situation is as follows:
df $= (n_1 - 1) + (n_2 - 1) = n_1 + n_2 - 2$.

KEY EXAMPLE

Two varieties of corn are being compared as to difference in maturation time. Ten plots of the first variety reach maturity in an average of 95 days with a standard deviation of 5.3 days, while eight plots of the second variety reach maturity in an average of 74 days with a standard deviation of 4.8 days. Determine a 95% confidence interval estimate for the difference in maturation time.

Answer:

$$n_1 = 10 \qquad n_2 = 8$$
$$\bar{x}_1 = 95 \qquad \bar{x}_2 = 74$$
$$s_1 = 5.3 \qquad s_2 = 4.8$$

$$\sigma_{\bar{x}-\bar{y}} = \sqrt{\frac{(10-1)(5.3)^2 + (8-1)(4.8)^2}{10+8-2}}\sqrt{\frac{1}{10}+\frac{1}{8}} = 2.41$$

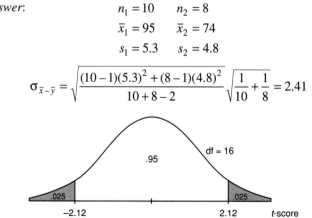

With df $= 10 + 8 - 2 = 16$, and .025 in each tail, the critical *t*-scores are ±2.12. The observed difference was $95 - 74 = 21$, so the confidence interval estimate is $21 \pm 2.12(2.41) = 21 \pm 5.1$. Thus, we can be 95% certain that the first variety of corn will take between 15.9 and 26.1 more days to reach maturity than will the second variety.

KEY EXAMPLE

A city councilmember claims that male and female officers wait equal times for promotion in the police department. A women's spokesperson believes women must wait longer than men. If five men waited 8, 7, 10, 5, and 7 years for promotion, while four women waited 9, 5, 12, and 8 years for promotion, what conclusion should be drawn at the 10% significance level?

Answer:

$$H_0: \mu_1 - \mu_2 = 0, \quad H_a: \mu_1 - \mu_2 < 0, \quad \alpha = .10$$

$$n_1 = 5, \qquad \bar{x}_1 = \frac{8 + 7 + 10 + 5 + 7}{5} = 7.4$$

$$s_1 = \sqrt{\frac{(8 - 7.4)^2 + (7 - 7.4)^2 + (10 - 7.4)^2 + (5 - 7.4)^2 + (7 - 7.4)^2}{5 - 1}} = 1.82$$

$$n_2 = 4, \qquad \bar{x}_2 = \frac{9 + 5 + 12 + 8}{4} = 8.5$$

$$s_2 = \sqrt{\frac{(9 - 8.5)^2 + (5 - 8.5)^2 + (12 - 8.5)^2 + (8 - 8.5)^2}{4 - 1}} = 2.89$$

$$\sigma_{\bar{x}-\bar{y}} = \sqrt{\frac{(5 - 1)(1.82)^2 + (4 - 1)(2.89)^2}{5 + 4 - 2}} \sqrt{\frac{1}{5} + \frac{1}{4}} = 1.57$$

With df $= 5 + 4 - 2 = 7$, and .10 in the tail, the critical t-score is -1.415. The critical difference is $-1.415(1.57) = -2.22$. Since the observed difference is only $7.4 - 8.5 = -1.1$, there is not sufficient evidence to dispute the councilmember's claim.

Key 42   Theme exercises with answers

OVERVIEW   *Sample questions of the type that might appear on homework assignments and tests are presented with answers.*

- In a sleeping laboratory experiment, 16 volunteers sleep an average of 7.4 hours with a standard deviation of 1.3 hours. What would be a 90% confidence interval estimate for the mean number of hours that people sleep at night?

  Answer: The standard deviation of the sample means $\sigma_{\bar{x}} = 1.3/\sqrt{16} = 0.325$. With $16 - 1 = 15$ degrees of freedom, and 5% in each tail, the appropriate t-scores are ± 1.753. Thus we can be 90% certain that the mean number of hours that people sleep at night is in the range $7.4 \pm 1.753(0.325) = 7.4 \pm 0.57$, or between 6.83 and 7.97 hours.

- Twenty recent graduates of a mathematics Ph.D. program were asked how many years they spent completing the program. Determine a 98% confidence interval estimate of the number of years to complete such a doctorate course of studies. The relevant totals are $\Sigma x = 86$ and $\Sigma x^2 = 371$.

  Answer: We calculate the sample mean $\bar{x} = 86/20 = 4.3$ and the sample standard deviation $s = \sqrt{(371 - 86^2/20)/19} = 0.251$. Then $\sigma_{\bar{x}} = 0.251/\sqrt{20} = 0.0562$. With $20 - 1 = 19$ degrees of freedom and 1% in each tail, the appropriate t-scores are ± 2.539. Therefore a 98% confidence interval estimate is given by $4.3 \pm 2.539(0.0562) = 4.3 \pm 0.14$, or between 4.16 and 4.44 years.

- A broker notes that the percentage gains in per-share value for stock in eight leisure-time companies during a particular 1-year period were 8.2, 9.5, 4.2, 10.0, 6.7, 6.6, 9.3, and 7.9. Find a 95% confidence interval estimate for the percentage gain in per-share value for leisure-time company stocks.

  Answer: The sample mean is

  $$\bar{x} = \frac{8.2 + 9.5 + 4.2 + 10.0 + 6.7 + 6.6 + 9.3 + 7.9}{8} = 7.8$$

and the standard deviation is

$$s = \sqrt{\frac{0.4^2 + 1.7^2 + 3.6^2 + 2.2^2 + 1.1^2 + 1.2^2 + 1.5^2 + 0.1^2}{7}} = 1.918$$

Also,

$$\sigma_{\bar{x}} = \frac{1.918}{\sqrt{8}} = 0.678$$

With df = 8 − 1 = 7, and 2.5% in each tail, the t-scores are ±2.365. A 95% confidence interval estimate is given by 7.8 ± 2.365 (0.678) = 7.8 ± 1.6, or between 6.2 and 9.4 percent.

- An IRS spokesperson claims that the average deduction for medical care is $1250. A taxpayer who believes that the real figure is lower samples 12 families and comes up with a mean of $934 and a standard deviation of $616. What conclusion should the taxpayer reach at a 5% significance level?

Answer:

$$H_0: \mu = 1250, \quad H_a: \mu < 1250, \quad \alpha = .05$$

$$\sigma_{\bar{x}} = \frac{616}{\sqrt{12}} = 177.8$$

With df = 12 − 1 = 11, and α = .05, the critical t-score is 1.796. The critical number of dollars is 1250 − 1.796(177.8) = 930.7. Since 934 > 930.7, at the 5% significance level the taxpayer should not reject the IRS spokesperson's claim.

- The weight of an aspirin tablet is 300 mg according to the bottle label. Should an FDA investigator reject the label if she weighs seven tablets and obtains weights of 299, 300, 305, 302, 299, 301, and 303 mg? Use a 2% significance level for this two-tailed test.

Answer:

$$H_0: \mu = 300, \quad H_a: \mu \neq 300, \quad \alpha = .02$$

$$\bar{x} = \frac{2109}{7} = 301.3$$

$$s = \sqrt{\frac{29.43}{6}} = 2.21$$

$$\sigma_{\bar{x}} = \frac{2.21}{\sqrt{7}} = 0.835$$

For $\alpha = .02$, we place a probability of .01 on each side, and note that df $= 7 - 1 = 6$ gives critical t-scores of ± 3.143. Critical weights are $300 \pm 3.143(0.835) = 300 \pm 2.62$, or 297.38 and 302.62 mg. Since 301.3 is in this range, at the 2% significance level the FDA investigator should not reject the label.

- The mean height of eight dwarf apple trees is 8.1 feet with a standard deviation of 2.3 feet, while the mean height of five dwarf cherry trees is 9.8 feet with a standard deviation of 2.8 feet. What is the 95% confidence interval estimate of the difference in heights between these two types of fruit trees?

Answer:

$$n_1 = 8 \qquad n_2 = 5$$
$$\bar{x}_1 = 8.1 \qquad \bar{x}_2 = 9.8$$
$$s_1 = 2.3 \qquad s_2 = 2.8$$

$$\sigma_{\bar{x}-\bar{y}} = \sqrt{\frac{(8-1)(2.3)^2 + (5-1)(2.8)^2}{8+5-2}}\sqrt{\frac{1}{8}+\frac{1}{5}} = 1.421$$

With df $= 8 + 5 - 2 = 11$, and .025 in each tail, the critical t-scores are ± 2.201. The observed difference was $8.1 - 9.8 = -1.7$, so the confidence interval estimate is $-1.7 \pm 2.201(1.421) = -1.7 \pm 3.13$. We can be 95% certain that the mean height of dwarf apple trees is between 1.43 feet greater and 4.83 feet less than the mean height of dwarf cherry trees.

- At a diplomatic party the Liberian ambassador comments that his embassy processes visa applications just as fast as the neighboring West African country of Sierra Leone. A reporter doubts this statement and the next day interviews eight applicants at the Liberian embassy (mean processing time 5 days with a standard deviation of 0.9 day) and ten applicants at the Sierra Leonean embassy (mean processing time 4.2 days with a standard deviation of 0.5 day). At the 10% significance level is the reporter correct to doubt the Liberian ambassador?

Answer:

$$H_0: \mu_1 - \mu_2 = 0, \quad H_a: \mu_1 - \mu_2 > 0, \quad \alpha = .10$$
$$n_1 = 8 \qquad n_2 = 10$$
$$\bar{x}_1 = 5 \qquad \bar{x}_2 = 4.2$$
$$s_1 = 0.9 \qquad s_2 = 0.5$$

$$\sigma_{\bar{x}-\bar{y}} = \sqrt{\frac{(8-1)(0.9)^2 + (10-1)(0.5)^2}{8+10-2}} \sqrt{\frac{1}{8} + \frac{1}{10}} = 0.334$$

With df $= 8 + 10 - 2 = 16$, and .10 in the tail, the critical t-score is 1.337. The critical difference is $1.337(0.334) = 0.45$ day. Since the observed difference is $5 - 4.2 = 0.8$ day, there is sufficient evidence for the reporter to doubt the Liberian ambassador's comment.

• A hospital exercise laboratory technician notes the resting pulse rates of five joggers to be: 60, 58, 59, 61, and 67, while the resting pulse rates of seven non-exercisers are: 83, 60, 75, 71, 91, 82, and 84. Establish a 90% confidence interval estimate for the difference in pulse rates between joggers and non-exercisers.

Answer:

$$\bar{x}_1 = \frac{60+58+59+61+67}{5} = 61$$

and

$$s_1 = \sqrt{\frac{1^2 + 3^2 + 2^2 + 0^2 + 6^2}{5-1}} = 3.54$$

$$\bar{x}_2 = \frac{83+60+75+71+91+82+84}{7} = 78$$

and

$$s_2 = \sqrt{\frac{5^2 + 18^2 + 3^2 + 7^2 + 13^2 + 4^2 + 6^2}{7-1}} = 10.23$$

$$\sigma_{\bar{x}-\bar{y}} = \sqrt{\frac{(5-1)(3.54)^2 + (7-1)(10.23)^2}{5+7-2}} \sqrt{\frac{1}{5} + \frac{1}{7}} = 4.82$$

With df $= 5 + 7 - 2 = 10$, and .05 in each tail, the critical t-scores are ± 1.812. The observed difference is $61 - 78 = -17$, so the confidence interval estimate is $-17 \pm 1.812(4.82) = -17 \pm 8.7$. Thus the technician can be 90% certain that joggers have an average resting pulse rate between 8.3 and 25.7 beats less than non-exercisers.

Theme 7 THE POPULATION PROPORTION

*T*hemes 5 and 6 dealt with estimates and tests for a population *mean*. Equally important in numerous applications are techniques and procedures involving a population *proportion*. For example, what proportion of registered voters support a proposed bond issue? What proportion of dentists recommend sugarless gum? What proportion of defense contracts are canceled because of cost overruns? What proportion of new houses sell during the first month on the market? Even when more refined measurements are possible, it may be sufficient to simply look at a proportion. For example, it may be necessary to know, not the mean salary of computer programmers, but rather just what proportion of the salaries are above $50,000. We may need to know, not the mean height of basketball players, but just the proportion of players above 6 feet 5 inches tall. And it may be necessary to know, not the mean number of new policies sold each week, but rather just the proportion of weeks for which at least ten new policies are sold.

Just as with means, it is usually impossible or impracticable to gather complete information about proportions. Thus we must use techniques to draw inferences about a population proportion when only a sample proportion is available. As we should now expect, results will involve confidence intervals and degrees of certainty; that is, we cannot make any inference about the population proportion with 100% certainty. However, with procedures very similar to those used to study means, and in fact making use of techniques developed for working with means, we will be able to draw inferences about proportions. Confidence interval estimates and hypothesis tests for both single-population proportions and differences of two proportions will be examined.

INDIVIDUAL KEYS IN THIS THEME

Key 43 The distribution of sample proportions

OVERVIEW *Whereas the mean is basically a quantitative measurement, the proportion is more of a qualitative approach. The interest is simply in the presence or absence of some attribute. We count the number of "yes" responses and form a proportion. For example, what proportion of drivers wear seat belts? What proportion of SCUD missiles can be intercepted? What proportion of new stereo sets have a certain defect? This separation of the population into "haves" and "have-nots" suggests that we can make use of our earlier work on binomial distributions.*

In this theme we are interested in estimating a population proportion π by considering a single sample proportion $\overline{p}$. This sample proportion is just one out of a whole universe of sample proportions, and to judge its significance we must know how sample proportions vary. Consider the set of proportions from all possible samples of a specified size n. It seems reasonable that these proportions will cluster around the population proportion, and that the larger the chosen sample size, the tighter will be the clustering.

How do we calculate the mean and variance of the set of sample proportions? Suppose the sample size is n, and the actual population proportion is π. From the keys on binomial distributions, we remember that the mean and standard deviation for the number of successes in a given sample are πn and $\sqrt{n\pi(1-\pi)}$, and for large n the complete distribution begins to look "normal."

We change to proportions by dividing each of these results by n:

$$\mu_{\overline{p}} = \frac{\pi n}{n} = \pi \qquad \text{and} \qquad \sigma_{\overline{p}} = \frac{\sqrt{n\pi(1-\pi)}}{n} = \sqrt{\frac{\pi(1-\pi)}{n}}$$

Thus the principle forming the basis of much of what we do in this theme is as follows:

Start with a population with a given proportion π. Take all samples of size n. Compute the proportion in each of these samples. Then:

- The set of all sample proportions will be approximately *normally* distributed.

- The *mean* of the set of sample proportions will equal π, the population proportion.

- The *standard deviation* $\sigma_{\bar{p}}$ of the set of sample proportions will be approximately equal to $\sqrt{\pi(1-\pi)/n}$.

KEY EXAMPLE

Suppose that 70% of all dialysis patients will survive at least five years. If 100 new dialysis patients are selected at random, what is the probability that the proportion surviving at least five years will exceed 80%?

Answer: The set of sample proportions is approximately normally distributed with mean .70 and standard deviation

$$\sigma_{\bar{p}} = \sqrt{\frac{(.7)(.3)}{100}} = 0.0458$$

With a z-score of $(.80 - .70)/0.0458 = 2.18$, the probability that our sample proportion exceeds 80% is $.5000 - .4854 = .0146$.

Key 44   Confidence interval estimate of
the proportion

OVERVIEW   *Using a measurement from a sample, we will never be able to say exactly what the population proportion is; rather, we will always say we have a certain confidence that the population proportion lies in a certain interval.*

In finding confidence interval estimates for the population proportion π, since π is unknown, how do we find

$$\sigma_{\bar{p}} = \sqrt{\frac{\pi(1-\pi)}{n}} \, ?$$

The reasonable procedure is to use the sample proportion $\bar{p}$:

$$\sigma_{\bar{p}} \approx \sqrt{\frac{\bar{p}(1-\bar{p})}{n}}$$

Finally, as we did with means, if we have a certain confidence that a sample proportion lies within a specified interval around the population proportion, then we have the same confidence that the population proportion lies within a specified interval about the sample proportion (the distance from Missoula to Whitefish is the same as the distance from Whitefish to Missoula).

KEY EXAMPLE

If 64% of a sample of 550 people leaving a shopping mall claim to have spent over \$25, determine a 99% confidence interval estimate for the proportion of shopping mall customers who spend over \$25.

Answer: Since $\bar{p} = .64$, the standard deviation of the set of sample proportions is

$$\sigma_{\bar{p}} = \sqrt{\frac{(.64)(.36)}{550}} = 0.0205$$

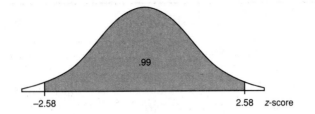

The 99% confidence interval estimate for the population proportion is .64 ± 2.58(0.0205) = .64 ± .053. Thus we are 99% certain that the proportion of shoppers spending over $25 is between .587 and .693.

KEY EXAMPLE

In a random sample of machine parts, 18 out of 225 were found to be damaged in shipment. Establish a 95% confidence interval estimate for the proportion of machine parts that are damaged in shipment.

Answer: The sample proportion is $\bar{p}$ = 18/225 = .08, and the standard deviation of the set of sample proportions is

$$\sigma_{\bar{p}} = \sqrt{\frac{(.08)(.92)}{225}} = 0.0181$$

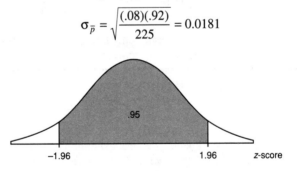

The 95% confidence interval estimate for the population proportion is .08 ± 1.96(0.0181) = .08 ± .035. Thus we are 95% certain that the proportion of machine parts damaged in shipment is between .045 and .115 .

Suppose there are 50,000 parts in the entire shipment. We can translate from proportions to actual numbers: .045(50,000) = 2250 and .115(50,000) = 5750, so we can be 95% confident that there are between 2250 and 5750 damaged parts in the whole shipment.

Some problems relating to the distribution of sample proportions involve one-sided intervals.

KEY EXAMPLE

An assembly-line quality check involves the following procedure. A sample of size 50 is randomly picked, and the machinery is shut down for repairs if the percentage of defective items in the sample is c percent or more. Find the value for c which results in a 90% chance that the machinery will be stopped if, on the average, it is producing 15% of defective items.

Answer: Since $\pi = .15$,

$$\sigma_{\bar{p}} = \sqrt{\frac{(.15)(.85)}{50}} = 0.0505$$

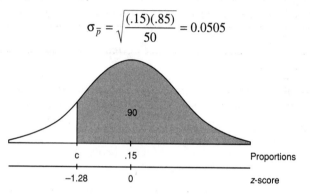

The z-score corresponding to the .90 probability is -1.28, so $c = .15 - 1.28(0.0505) = .085$ or 8.5%.

Key 45   Selecting a sample size

OVERVIEW   *One important consideration in setting up surveys is the choice of sample size. To obtain a smaller, more precise interval estimate of the population proportion, we must either decrease the degree of confidence or increase the sample size. Similarly, if we want to increase the degree of confidence, we may either accept a wider interval estimate or increase the sample size. Choosing a larger sample size seems desirable; in the real world, however, this decision involves time and cost considerations.*

In setting up a survey to obtain a confidence interval estimate of the population proportion, what should we use for $\sigma_{\bar{p}}$? It can be shown that $\sqrt{\pi(1-\pi)}$ is no larger than .5. Thus $\sqrt{[\pi(1-\pi)]/n}$ is at most $.5/\sqrt{n}$. We make use of this fact in determining sample sizes in problems such as the following.

KEY EXAMPLE

An EPA investigator wants to know the proportion of fish that are inedible because of chemical pollution downstream of an offending factory. If the answer needs to be within ±.03 at the 96% confidence level, how many fish should be in the sample?

Answer:

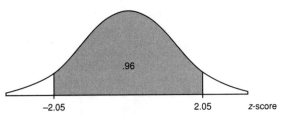

We want $2.05\sigma_{\bar{p}} \le .03$. By the statement given above, $\sigma_{\bar{p}}$ is at most $.5/\sqrt{n}$, so it is sufficient to consider $2.05(.5/\sqrt{n}) \le .03$. Algebraically, we get $\sqrt{n} \ge 2.05(.5)/.03 = 34.17$, and so $n \ge 1167.4$.

Therefore, choosing a sample of 1168 fish will give the inedible proportion to within ±.03 at the 96% level.

Note that the accuracy of the above estimate does *not* depend on what fraction of the whole population we have sampled. What is critical is the absolute size of the sample.

How does the sample size change the accuracy?

Is some minimal value of n necessary for these procedures to be meaningful? Since we are using the normal approximation to the binomial, as a minimum we should have finally that both np and $n(1 - p)$ are at least 5 (see Key 26).

KEY EXAMPLE

A study is to be undertaken to determine the proportion of industry executives who believe that workers' pay should be based on individual performance. How many executives should be interviewed if an estimate is desired at the 99% confidence level to within ±0.6? To within ±.03? To within ±.02?

Answer: Algebraically, $2.58(.5/\sqrt{n}) \leq .06$ gives $\sqrt{n} \geq 2.58(.5)/.06 = 21.5$, and so $n \geq 462.25$. Similarly, $2.58(.5/\sqrt{n}) \leq .03$ gives $\sqrt{n} \geq 2.58(.5)/.03 = 43$, and so $n \geq 1849$. Finally, $2.58(.5/\sqrt{n}) \leq .02$ gives $\sqrt{n} \geq 2.58(.5)/.02 = 64.5$, and so $n \geq 4160.25$. Thus 463, 1849, or 4161 executives should be interviewed depending upon the accuracy desired.

Note that to cut the interval estimate in half (from ±.06 to ±.03), we would have to increase the sample size fourfold (from 462.25 to 1849). To cut the interval estimate to a third (from ±.06 to ±.02), we would have to increase the same size ninefold (from 462.25 to 4160.25).

More generally, if we want to divide the interval estimate by d without effecting the confidence level, we must increase the sample size by a multiple of d^2.

Key 46 Hypothesis test of the proportion

OVERVIEW *Closely related to the problem of estimating a population proportion is the problem of testing a hypothesis about a population proportion. For example, a travel agency might determine an interval estimate for the proportion of sunny days in the Virgin Islands or, alternatively, might test a tourist bureau's claim about the proportion of sunny days. A major stockholder might ascertain an interval estimate for the proportion of successful contract bids or, alternatively, could test a company spokesperson's claim about the proportion of successful bids. A social scientist could find an interval estimate for the proportion of homeless children who attend school or, alternatively, might test a school board member's claim about the proportion of such children who are still able to go to classes. In each of the above, the researcher must decide whether the interest lies in an interval estimate of a population proportion or in a hypothesis test of a claimed proportion.*

The general testing procedure is very similar to that developed for testing a claim about a population mean. There is a specific hypothesis to be tested, called the *null hypothesis*, which is stated in the form of an equality statement about the population proportion (for example, H_0: $\pi = .37$). There is the *alternative hypothesis,* stated in the form of an inequality (for example, H_a: $\pi < .37$ or H_a: $\pi > .37$ or H_a: $\pi \neq .37$). The null hypothesis cannot be proved incorrect with absolute certainty; rather, it may be shown to be improbable. The testing procedure involves picking a sample and comparing the sample proportion $\bar{p}$ to the claimed population proportion π. A *critical value,* or critical values, c is (are) chosen to gauge the significance of the sample statistic. If the observed $\bar{p}$ is further from the claimed proportion π than is the critical value, then we say that there is sufficient evidence to reject the null hypothesis. For example, if H_0: $\pi < .37$ and if $c = .33$, then a sample $\bar{p} = .35$ would not be sufficient evidence to reject H_0.

In practice, one chooses an acceptable α-*risk*, or probability of committing a *Type I error* and mistakenly rejecting a true null hypothesis. This α-risk, also called the *significance level* of the test, is used to determine the critical value(s). The strength of the sample statistic $\bar{p}$ can further be gauged through its associated p-value, which gives the smallest value of α for which H_0 would be rejected. Finally, one should consider the possibility of a mistaken failure to reject a false null hypothesis. This is called a *Type II error* and has associated probability β. There is a different β-value for each possible correct value for the population parameter π.

Key 47 Critical values, α-risks, and *p*-values

OVERVIEW *We test a null hypothesis by picking a sample and comparing the sample proportion to the claimed population proportion. Critical values are chosen to gauge the significance of our sample statistic. For a given critical value, we can sketch the distribution of sample proportions around the claimed population proportion, and then note the region corresponding to the probability of a Type I error or α-risk. Alternatively, starting with an acceptable α-risk, we can calculate the corresponding critical value. If the sample statistic is far enough from the claimed population parameter, we say there is sufficient evidence to reject the null hypothesis. The* p-*value of the test is the smallest value of* α *for which the null hypothesis would be rejected.*

KEY EXAMPLE

A local restaurant owner claims that only 15% of visiting tourists stay for more than 2 days. A chamber of commerce volunteer is sure that the real percentage is higher. He plans to survey 100 tourists and intends to speak up if at least 18 of the tourists stay for over 2 days. What is the probability of a Type I error?

Answer:

$$H_0: \pi = .15, \quad H_a: \pi > .15$$

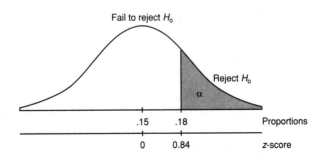

Using the claimed 15%, we calculate the standard deviation of sample proportions to be

$$\sigma_{\bar{p}} = \sqrt{\frac{(.15)(.85)}{100}} = 0.0357$$

The critical proportion is $c = 18/100 = .18$, so the critical z-score is $(.18 - .15)/0.0357 = 0.84$. Thus $\alpha = .5000 - .2995 = .2005$. The test, as set up by the volunteer, has a 20.05% chance of mistakenly rejecting a true null hypothesis.

KEY EXAMPLE

A union spokesperson claims that 75% of the union members support a strike if their basic demands are not met. A company negotiator believes the true percentage is lower and runs a hypothesis test at the 10% significance level. What is the conclusion if 87 of 125 union members say they will strike?

Answer:

$$H_0: \pi = .75, \quad H_a: \pi < .75$$

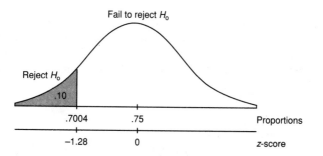

We use the claimed proportion to calculate the standard deviation of the sample proportions.

$$\sigma_{\bar{p}} = \sqrt{\frac{(.75)(.25)}{125}} = 0.03873$$

With $\alpha = .10$ the critical z-score is -1.28, so the critical proportion is $.75 - 1.28(0.03873) = .7004$. The observed sample proportion is $\bar{p} = 87/125 = .696$. Since $.696 < .7004$, there is sufficient evidence to reject H_0 at the 10% significance level. The company negotiator should challenge the union claim.

To measure the strength of the disagreement between the sample proportion and the claimed proportion, we calculate the *p*-value (also called the *attained significance level*). The *z*-score of .696 is (.696 − .75)/0.03873 = −1.39 with a resulting *p*-value of .5 − .4177 = .0823. We note that there is not sufficient evidence to reject H_0 at the 5%, or even the 8%, significance level.

KEY EXAMPLE

A cancer research group surveys 500 women over 40 years old to test the hypothesis that 28% of this age group have regularly scheduled mammograms. Should the hypothesis be rejected at the 5% significance level if 151 of the women respond affirmatively?

Answer: Since no suspicion has been voiced that the 28% claim is low or high, we run a two-sided (also called two-tailed) test.

$$H_0: \pi = .28, \quad H_a: \pi \neq .28, \quad \alpha = .05$$

$$\sigma_{\bar{p}} = \sqrt{\frac{(.28)(.72)}{500}} = 0.0201$$

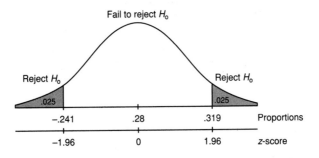

With $\alpha/2 = .025$ in each tail, the critical *z*-scores are ±1.96, and the critical proportions are .28 ± 1.96(0.0201) = .241 and .319. The observed $\bar{p} = 151/500 = .302$. Since .302 is *not* outside the critical values, there is *not* sufficient evidence to reject H_0; that is, the cancer research group should not dispute the 28% claim.

What is the *p*-value?

Answer: The *z*-score of .302 is (.302 − .28)/0.0201 = 1.09, which corresponds to a probability of .5000 − .3621 = .1379. Doubling (because the test is two-sided), we obtain a *p*-value of 2(.1379) = .2758. Thus, for example, the null hypothesis should not be rejected even if α were a relatively large .25.

Key 48 Type II errors

OVERVIEW *Why not always choose α to be extremely small so as to eliminate the possibility of mistakenly reject-ing a correct null hypothesis? The difficulty is that this choice would simultaneously increase the chance of never rejecting the null hypothesis even if it were far from true. Thus we see that, for a more nearly complete picture, we must also calculate the probability β of mistakenly failing to reject a false null hypothesis. As was the case with means, we shall see that there is a different β-value for each possible correct value for the population proportion π. The operat-ing characteristic curve, or OC curve, a graphical display of β-values, is often given in the real-life analysis of a hypoth-esis test.*

KEY EXAMPLE

A soft-drink manufacturer received a 9% share of the market this past year. The marketing research department plans a telephone survey of 3000 households. If less than 8% indicate they will buy the company's product, the research department will conclude that the market share has dropped and will order special new promotions. What is the proba-bility of a Type I error?

Answer:

$$H_0: \pi = .09, \quad H_a: \pi < .09$$

$$\sigma_{\bar{p}} = \sqrt{\frac{(.09)(.91)}{3000}} = 0.005225$$

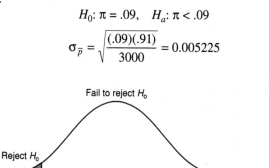

The z-score of .08 is $(.08 - .09)/0.005225 = -1.91$, so $\alpha = .5 - .4719 = .0281$.

What is the probability of a Type II error if the true market share is .085? In other words, if the market share really has dropped (to 8.5%), what is the probability that the research department will mistakenly fail to reject the 9% null hypothesis?

Answer: The z-score of .08 now is $(.08 - .085)/0.005225 = -0.96$. Thus $\beta = .5000 + .3315 = .8315$.

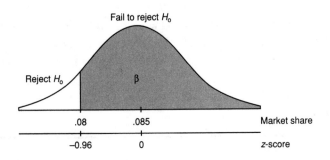

What if the true market share is .075?

Answer: The z-score of .08 now is $(.08 - .075)/0.005225 = 0.96$. Thus $\beta = .5000 - .3315 = .1685$.

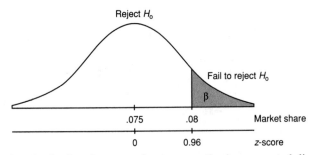

Note that the further the true value is away (in the suspected direction) from the claimed null hypothesis, the smaller the probability is of failing to reject the false claim.

KEY EXAMPLE

A building inspector believes that the percentage of new construction with serious code violations may be even greater than the previously claimed 7%. A hypothesis test is planned on 200 new homes at the 10% significance level. What is the β-value if the true percentage of new constructions with serious violations is 9%? Is 11%? Is 13%?

Answer:

$$H_0: \pi = .07, \quad H_a: \pi > .07, \quad \alpha = .10$$

$$\sigma_{\bar{p}} = \sqrt{\frac{(.07)(.93)}{200}} = 0.018$$

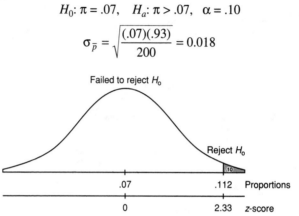

The 10% significance level gives a critical z-score of 2.33 and a critical proportion of .07 + 2.33(0.018) = .112.

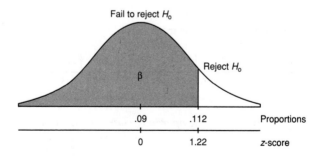

If the true percentage is 9%, the z-score of .112 is (.112 – .09)/0.018 = 1.22. Then β = .5000 + .3888 = .8888.

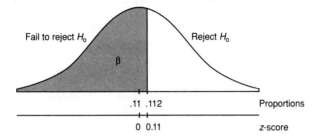

If the true percentage is 11%, the z-score of .112 is $(.112 - .11)/0.018 = 0.11$. Then $\beta = .5000 + .0438 = .5438$.

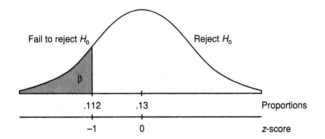

If the true percentage is 13%, the z-score of .112 is $(.112 - .13)/0.018 = -1$. Then $\beta = .5000 - .3413 = .1587$.

A table of these values (and a few more) for β is as follows:

True π	.08	.09	.10	.11	.12	.13	.14	.15
β	.9625	.8888	.7486	.5438	.3300	.1587	.0594	.0174

The resulting graph is called the *operating characteristic (OC) curve*.

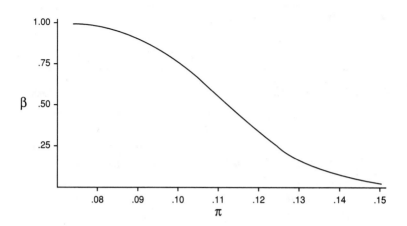

If β is the probability of failing to reject a false null hypothesis, then $1 - \beta$ is the probability of rejecting the false null hypothesis. The *power* of a hypothesis test is the probability that a Type II error is not committed, and the graph of $1 - \beta$ is called the *power curve*.

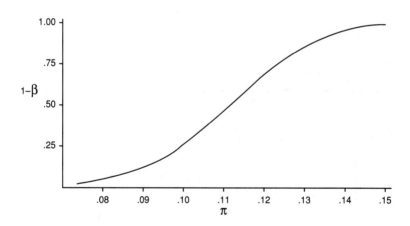

Both the operating characteristic curve and the related power curve help one to gauge the effectiveness of a hypothesis test.

Key 49   Confidence interval estimate for the difference of two proportions

OVERVIEW   *Numerous important and interesting applications of statistics involve the comparison of two population proportions. For example, is the proportion of satisfied purchasers of American automobiles greater than that for buyers of Japanese cars? How does the percentage of surgeons recommending a new cancer treatment compare with the corresponding percentage of oncologists? What can be said about the difference between the proportion of single parents who are on welfare and the proportion of two-parent families on welfare?*

Often it is clear that some difference exists between two population proportions, and we would like to give a numerical measure of the difference. Using samples, we cannot give this difference *exactly*; however, we can say, with a specified degree of confidence, that the difference lies in a certain interval. We follow the same procedure as set forth in Key 44, this time using

$$\pi_1 - \pi_2, \qquad \sqrt{\frac{\pi_1(1-\pi_1)}{n_1} + \frac{\pi_2(1-\pi_2)}{n_2}}, \qquad \text{and} \qquad \bar{p}_1 - \bar{p}_2$$

in place of

$$\pi, \qquad \sqrt{\frac{\pi(1-\pi)}{n}}, \qquad \text{and} \qquad \bar{p}$$

As in the preceding keys on proportions, we are using the normal to estimate the binomial and so will assume that

$$n_1\pi_1, \quad n_1(1-\pi_1), \quad n_2\pi_2, \quad \text{and} \quad n_2(1-\pi_2)$$

are all at least 5.

KEY EXAMPLE

Suppose that 84% of a sample of 125 nurses working 7 A.M. to 3 P.M. shifts in city hospitals express positive job satisfaction, while only 72% of a sample of 150 nurses on 11 P.M. to 7 A.M. shifts express similar fulfillment. Establish a 90% confidence interval estimate for the difference.

Answer:

$$n_1 = 125, \quad n_2 = 150$$

$$\bar{p}_1 = .84, \quad \bar{p}_2 = .72$$

$$\sigma_d = \sqrt{\frac{(.84)(.16)}{125} + \frac{(.72)(.28)}{150}} = 0.0492$$

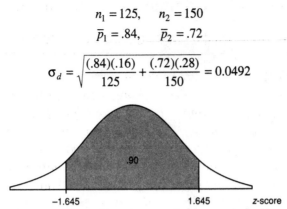

The observed difference is $.84 - .72 = .12$, and the critical z-scores are ± 1.645. The confidence interval estimate is $.12 \pm 1.645(0.0492) = .12 \pm .081$. We can be 90% certain that the proportion of satisfied nurses on 7 to 3 shifts is between .039 and .201 higher than for those on 11 to 7 shifts.

KEY EXAMPLE

A grocery store manager notes that, in a sample of 85 people going through the "under 7 items" checkout line, only 10 paid with checks; whereas, in a sample of 92 customers passing through the regular line, 37 paid with checks. Find a 95% confidence interval estimate for the difference between the proportions of customers going through the two different lines who use checks.

Answer:

$$n_1 = 85, \quad n_2 = 92$$

$$\bar{p}_1 = \frac{10}{85} = .118, \quad \bar{p}_2 = \frac{37}{92} = .402$$

$$\sigma_d = \sqrt{\frac{(.118)(.882)}{85} + \frac{(.402)(.598)}{92}} = 0.0619$$

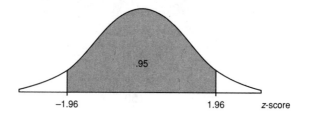

.95

−1.96   1.96   *z*-score

The observed difference is $.118 - .402 = -.284$, and the critical z-scores are ± 1.96. Thus, (see also Key 50) the confidence interval estimate is $-.284 \pm 1.96(0.0619) = -.284 \pm .121$. The manager can be 90% sure that the proportion of customers passing through the "under 7 items" line who use checks is between .163 and .405 lower than the proportion going through the regular line who use checks.

Setting up experiments or surveys involves many considerations, one of which is *sample size*. Generally, if we want smaller, more precise interval estimates, we either decrease the degree of confidence or increase the sample size. Similarly, if we want to increase the degree of confidence, we may either accept a wider interval or again increase the sample size.

In Key 45 we noted that $\sqrt{\pi(1 - \pi)}$ is at most .5. Thus (see also Key 50)

$$\sqrt{\pi(1 - \pi)\left(\frac{1}{n_1} + \frac{1}{n_2}\right)} \le (.5)\sqrt{\frac{1}{n_1} + \frac{1}{n_2}}$$

Now, if we simplify by insisting that $n_1 = n_2 = n$, the above expression reduces as follows:

$$(.5)\sqrt{\frac{1}{n} + \frac{1}{n}} = (.5)\sqrt{\frac{2}{n}} = \frac{.5\sqrt{2}}{\sqrt{n}}$$

KEY EXAMPLE

A pollster wants to determine the difference between the proportions of high-income voters and of low-income voters who support a decrease in capital gains taxes. If the answer needs to be known to within $\pm.02$ at the 95% confidence level, what size samples should be taken?

Answer: Assuming we will pick the same size samples for the two sample proportions, we have

$$\sigma_d \le \frac{.5\sqrt{2}}{\sqrt{n}} \quad \text{and} \quad 1.96\sigma_d \le .02$$

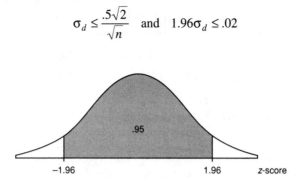

Thus, $1.96(.5)\sqrt{2}/\sqrt{n} \le .02$, and algebraically we find that

$$\sqrt{n} \ge \frac{1.96(.5)\sqrt{2}}{.02} = 69.3$$

Therefore, $n \ge 69.3^2 = 4802.5$, and the pollster should use 4803 people for each sample.

Key 50 Hypothesis test for the difference of two proportions

OVERVIEW *To compare proportions from two different populations, we consider a sample from each population and note the difference between the sample proportions. The strength of this difference can be measured by calculating the* p-*value.*

We consider problems for which the null hypothesis states that the population proportions are equal, or, equivalently, that their difference is 0:

$$H_0: \pi_1 - \pi_2 = 0$$

The alternative hypothesis is then

$$H_a: \pi_1 - \pi_2 < 0, \quad H_a: \pi_1 - \pi_2 > 0, \quad \text{or} \quad H_a: \pi_1 - \pi_2 \neq 0$$

The first two possibilities lead to one-sided (one-tailed) tests, while the third possibility leads to two-sided (two-tailed) tests.

Since the null hypothesis is that $\pi_1 = \pi_2$, we call this common value π, and use it to calculate σ_d:

$$\sigma_d = \sqrt{\frac{\pi(1-\pi)}{n_1} + \frac{\pi(1-\pi)}{n_2}} = \sqrt{\pi(1-\pi)\left(\frac{1}{n_1} + \frac{1}{n_2}\right)}$$

In practice, if

$$\bar{p}_1 = \frac{x_1}{n_1} \quad \text{and} \quad \bar{p}_2 = \frac{x_2}{n_2}$$

we use

$$\bar{p} = \frac{x_1 + x_2}{n_1 + n_2}$$

as an estimate for π in calculating σ_d.

KEY EXAMPLE

Suppose, early in an election campaign, a telephone poll of 800 registered voters shows 460 in favor of a particular candidate. Just before election day, a second poll shows 520 out of 1000 registered voters expressing this preference. At the 10% significance level is there sufficient evidence that the candidate's popularity has decreased?

Answer: $H_0: \pi_1 - \pi_2 = 0, \quad H_a: \pi_1 - \pi_2 > 0, \quad \alpha = .10$

$$\bar{p}_1 = \frac{460}{800} = .575, \quad \bar{p}_2 = \frac{520}{1000} = .520, \quad \bar{p} = \frac{460 + 520}{800 + 1000} = .544$$

$$\sigma_d = \sqrt{(.544)(.456)\left(\frac{1}{800} + \frac{1}{1000}\right)} = 0.0236$$

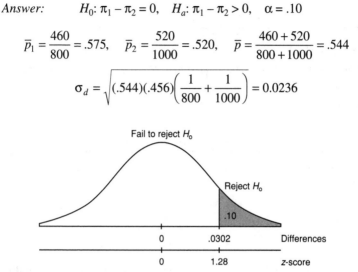

With $\alpha = .10$, the critical z-score is 1.28, so the critical difference is $0 + 1.28(0.0236) = .0302$. The observed difference is $.575 - .520 = .055$. Since $.055 > .0302$, we conclude that at the 10% significance level the candidate's popularity *has* dropped.

The strength of this drop can be further measured by calculating the p-value. The z-score of .055 is $(.055 - 0)/0.0236 = 2.33$, so the p-value is $.5000 - .4901 = .0099$. Thus there is sufficient evidence of a drop in popularity even at the 1% significance level.

KEY EXAMPLE

An automobile manufacturer tries two distinct assembly procedures. In a sample of 350 cars coming off the line using the first procedure, there are 28 with major defects, while a sample of 500 autos from the second line shows 32 with defects. Is the difference significant at the 6% significance level?

Answer: Since there is no mention that one of the procedures is believed to be better or worse than the other, this is a two-sided test.

$$H_0: \pi_1 - \pi_2 = 0, \quad H_a: \pi_1 - \pi_2 \neq 0, \quad \alpha = .06$$

$$\bar{p}_1 = \frac{28}{350} = .080, \quad \bar{p}_2 = \frac{32}{500} = .064, \quad \bar{p} = \frac{28 + 32}{350 + 500} = .0706$$

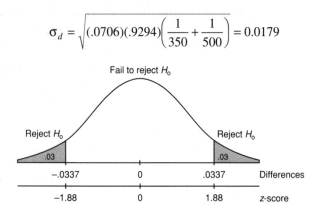

$$\sigma_d = \sqrt{(.0706)(.9294)\left(\frac{1}{350} + \frac{1}{500}\right)} = 0.0179$$

With $\alpha = .06$, we put .03 in each tail, and note that the critical z-scores are ± 1.88. The critical differences are $0 \pm 1.88(0.0179) = \pm.0337$. The observed difference is $.080 - .064 = .016$. Since .016 is between the two critical values, we conclude that the observed difference is not significant at the 6% level.

What is the p-value? The z-score of .016 is $(0 - .016)/0.0179 = 0.89$, which corresponds to a tail probability of $.5000 - .3133 = .1867$. Doubling because the test is two-sided results in a p-value of $2(.1867) = .3734$. Thus the smallest significance level for which the observed difference is significant is a very large 37.34%.

Key 51   Theme exercises with answers

OVERVIEW   *Sample questions of the type that might appear on homework assignments and tests are presented with answers.*

- In one study, 18% of 100 people with migraine headaches experienced substantial symptomatic relief after taking a placebo. Establish a 90% confidence interval estimate of the percentage of migraine sufferers who can be helped by placebos.

  Answer: Since $\bar{p} = .18$, the standard deviation of the set of sample proportions is

  $$\sigma_{\bar{p}} = \sqrt{\frac{(.18)(.82)}{100}} = 0.03842$$

  The 90% confidence interval estimate for the population proportion is $.18 \pm 1.645(0.03842) = .18 \pm .063$. Thus we are 90% certain that the proportion of migraine sufferers who can be helped by placebos is between .117 and .243.

- In a random sample of 75 clothing purchases returned for refunds, the buyers claimed that 47 were brought back because of improper fit. Construct a 94% confidence interval estimate for the proportion of clothing returns that are blamed on fit.

  Answer: The sample proportion is $\bar{p} = 47/75 = .627$, and the standard deviation of the set of sample proportions is

  $$\sigma_{\bar{p}} = \sqrt{\frac{(.627)(.373)}{75}} = 0.0558$$

  The 94% confidence interval estimate for the population proportion is $.627 \pm 1.88(.0558) = .627 \pm .105$. Thus we are 94% certain that between .522 and .732 of the clothing returns are blamed on poor fit.

- A Department of Labor survey of 6230 unemployed adults classified people by marital status, sex, and race. The raw numbers are as follows:

	White, 16 Years and Over			Nonwhite, 16 Years and Over		
	Married	Widow/Div.	Single	Married	Window/Div.	Single
Men	1090	337	1168	266	135	503
Women	952	423	632	189	186	349

Find a 90% confidence interval estimate for the proportion of unemployed men who are married.

Answer: Totaling the first row across, we find that there were 3499 men in the survey, and we note that 1090 + 266 = 1356 of these were married. Therefore,

$$\bar{p} = \frac{1356}{3499} = .3875 \quad \text{and} \quad \sigma_{\bar{p}} = \sqrt{\frac{(.3875)(.6125)}{3499}} = 0.008236$$

Thus the 90% confidence interval estimate is .3875 ± 1.645(0.008236) = .3875 ± .0135.

Find a 98% confidence interval estimate of the proportion of unemployed singles who are women.

Answer: There were 1168 + 632 + 503 + 349 = 2652 singles in the survey, and of these 632 + 349 = 981 were women. Therefore,

$$\bar{p} = \frac{981}{2652} = 0.3699 \quad \text{and} \quad \sigma_{\bar{p}} = \sqrt{\frac{(.3699)(.6301)}{2652}} = 0.009375$$

The 98% confidence interval estimate is .3699 ± 2.33(0.009375) = .3699 ± .0218.

- A telephone survey of 1000 adults was taken shortly after the United States began bombing Iraq. If 832 voiced their support for this action, with what confidence can it be asserted that 83.2% ± 3% of the adult population supported the decision to go to war?

$$\bar{p} = \frac{832}{1000} = .832 \quad \text{and} \quad \sigma_{\bar{p}} = \sqrt{\frac{(.832)(.168)}{1000}} = 0.0118$$

The relevant z-scores are ±.03/0.0118 = ±2.54. Table A gives a probability of .4945, and so our answer is 2(.4945) = .9890. In other words, 83.2% ± 3% is a 98.90% confidence interval estimate for adult support of the war decision.

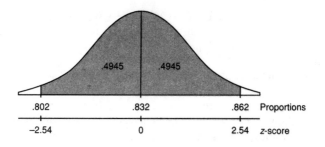

If the adult U.S. population is 191,000,000, estimate the actual numerical support.

Answer: Since .802(191,000,000) ≈ 153,000,000, while .862(191,000,000) ≈ 165,000,000, we can be 98.90% sure that between 153 and 165 million adults supported the initial bombing decision.

- The American Medical Association wishes to determine the percentage of obstetricians who are considering leaving the profession because of the rapidly increasing number of lawsuits against obstetricians. How large a sample should be taken to find the answer to within ±2% at the 95% confidence level?

Answer: We want $1.96\sigma_{\bar{p}} \leq .02$. Since $\sigma_{\bar{p}}$ is at most $.5/\sqrt{n}$, it is sufficient to consider $1.96(.5/\sqrt{n}) \leq .02$. Algebraically, we get $\sqrt{n} \geq 1.96(.5)/.02 = 49$, and so $n \geq 2401$. Thus interviewing 2401 obstetricians will give the required proportion to within ±.02 at the 95% level.

- It is believed that 60% of all whiskey drinkers can tell the difference between Rot Gut and Northern Comfort. A hypothesis test is run on 50 pub customers, and 35 are able to distinguish between the whiskeys. Is this sufficient evidence at the 5% significance level to call into question the 60% claim?

Answer:
$$H_0\text{: } \pi = .60, \quad H_a\text{: } \pi \neq .60, \quad \alpha = .05$$

We use the claimed 60% to calculate the standard deviation of sample proportions to be
$$\sigma_{\bar{p}} = \sqrt{\frac{(.60)(.40)}{50}} = 0.0693$$

With $\alpha/2 = .025$ in each tail, the critical z-scores are ±1.96, and the critical proportions are .60 ± 1.96(0.0693) = .464 and .736. The observed $\bar{p} = 35/50 = .70$. Since .70 is not outside the critical values, there is *not* sufficient evidence to reject H_0. In other words, at the 5% significance level, the 60% claim should not be disputed.

- A television producer knows that, unless a new show is watched by at least 25% of the possible audience, it will be canceled. Suppose a quick survey finds only 21 out of 100 viewers tuned in to the show. Is this sufficient evidence at the 10% significance level that the show will be canceled? Determine the *p*-value.

Answer:

$$H_0: \pi = .25, \quad H_a: \pi < .25, \quad \alpha = .10$$

$$\sigma_{\bar{p}} = \sqrt{\frac{(.25)(.75)}{100}} = 0.0433$$

With $\alpha = .10$ the critical *z*-value is 2.33 and the critical proportion is $.25 - 2.33(0.0433) = .149$. Since $\bar{p} = 21/100 = .21$ and $.21 > .149$, there is *not* sufficient evidence to cancel the new show.

The *z*-score of .21 is $(.21 - .25)/0.0433 = -0.924$ with a resulting *p*-value of $.5 - .3212 = .1788$. Thus the show would not be canceled even at the 17% significance level, but it could be at the 18% level.

- An airline claims that 92% of its flights leave on schedule, but an FAA investigator believes the true figure is lower. He decides that 125 flights will be checked at the 5% significance level. What is the probability of a Type II error if the true percentage is 90%?

Answer:

$$H_0: \pi = .92, \quad H_a: \pi < .92, \quad \alpha = .05,$$

$$\sigma_{\bar{p}} = \sqrt{\frac{(.92)(.08)}{125}} = 0.0243$$

With $\alpha = .05$ the critical *z*-score is 1.645, and the critical proportion is $.92 - 1.645(0.0243) = .880$. If the true proportion of on-schedule flights is .90, then the *z*-score of .880 is $(.880 - .90)/0.0243 = 0.82$, and $\beta = .5 + .2939 = .7939$.

- A survey of 5000 medical students compared the career goals of men and women.

Career Goal

	Surgery	Gynecology	Pediatrics	Psychiatry	Other
Men	312	520	472	610	1026
Women	128	350	391	400	791

Establish a 95% confidence interval estimate for the difference between the proportions of men and women intending to become surgeons.

Answer:

$$\bar{p}_1 = \frac{312}{312 + 520 + 472 + 610 + 1026} = \frac{312}{2940} = .106$$

$$\bar{p}_2 = \frac{128}{128 + 350 + 391 + 400 + 791} = \frac{128}{2060} = .062$$

$$\sigma_d = \sqrt{\frac{(.106)(.894)}{2940} + \frac{(.062)(.938)}{2060}} = 0.00778$$

The observed difference is $.106 - .062 = .044$, so the confidence interval estimate is $.044 \pm 1.96(0.00778) = .044 \pm .015$. Therefore we can be 95% certain that the proportion of men heading toward surgery is between .029 and .059 higher than the proportion of women.

Is the observed difference between the proportions of men and women hoping to become gynecologists significant at the 2% significance level?

Answer:

$$H_0: \pi_1 - \pi_2 = 0, \quad H_a: \pi_1 - \pi_2 \neq 0, \quad \alpha = .02$$

$$\bar{p}_1 = \frac{520}{2940} = .177, \quad \bar{p}_2 = \frac{350}{2060} = .170, \quad \bar{p} = \frac{520 + 350}{5000} = .174$$

$$\sigma_d = \sqrt{(.174)(.826)\left(\frac{1}{2940} + \frac{1}{2060}\right)} = 0.0109$$

The critical z-scores are ± 2.33, so the critical differences are $0 \pm 2.33(0.0109) = \pm.025$. The observed difference is $.177 - .170 =$

.007. Since .007 is between −.025 and .025, there is *not* sufficient evidence to reject H_0. The observed difference between the proportions of men and women looking toward gynecology is not significant at the 2% level.

Is there sufficient evidence at the 0.5% significance level that a higher proportion of women want to become pediatricians than do men?

Answer:

$$H_0: \pi_1 - \pi_2 = 0, \quad H_a: \pi_1 - \pi_2 > 0, \quad \alpha = .005,$$

$$\bar{p}_1 = \frac{391}{2060} = .190, \quad \bar{p}_2 = \frac{472}{2940} = .161, \quad \bar{p} = \frac{391 + 472}{5000} = .173$$

$$\sigma_d = \sqrt{(.173)(.827)\left(\frac{1}{2060} + \frac{1}{2940}\right)} = 0.0109$$

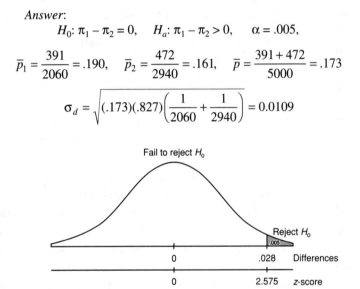

The critical z-score is 2.575, so the critical difference is $0 + 2.575(0.0109) = .028$. The observed difference is $.190 - .161 = .029$. Since $.029 > .028$, there is sufficient evidence to reject H_0. At the 0.5% significance level there is evidence that a greater proportion of women than men plan to become pediatricians.

Theme 8   CHI-SQUARE ANALYSIS

*I*n this theme we consider two types of problems involving similar analyses. The first concerns whether or not some observed distributional outcome fits some previously specified pattern. Perfect fits very rarely exist, and we must develop a procedure for measuring the significance of a loose fit. The second type of problem concerns whether two variables are independent or have some relationship. Here we use the same kind of measurement as above to judge the significance of a loose fit with the theoretical pattern based on independence. The kinds and the variety of problems we will be considering are indicated by the following examples.

A geneticist might test whether inherited traits can be explained by a *binomial distribution.* A physicist might look to a *Poisson distribution* to further understand alpha-particle emissions. A psychologist might study human intelligence patterns in terms of a *normal distribution.* In these and many other applications, the question arises as to how well observed data fit the pattern expected from some specified distribution.

A pollster might look at whether or not voter support for a candidate is *independent* of the voters' ethnic backgrounds. An efficiency expert might look at whether or not the likelihood of an accident is *independent* of which shift an employee works on. A psychologist might test whether or not certain mental abnormalities are *independent* of socioeconomic background.

As can be seen, the procedure described in the following keys has a wide range of applications. **Chi-square analysis** is one of the most useful statistical techniques, both where other tests are not applicable, and where other tests may be unnecessarily complicated.

Key 52 Chi-square calculation

OVERVIEW *A critical question in statistics is whether or not some observed pattern of data fits some given distribution. Since a perfect fit cannot be expected, we must be able to look at discrepancies and make a judgment as to the "goodness of fit." The best information is obtained by squaring the discrepancy values and then appropriately **weighting** each difference. Such weighting is accomplished by dividing each difference by the expected value. The sum of these weighted differences or discrepancies is called **chi-square,** and is denoted as χ^2 (χ is the lower-case Greek letter chi):*

$$\chi^2 = \sum \frac{(\text{obs.} - \text{exp.})^2}{\text{exp.}}$$

KEY EXAMPLE

Census figures show that a city has 64% white, 25% black, and 11% Hispanic residents. A sampling of 350 new city employees shows 243 white, 80 black, and 27 Hispanic. Is the city hiring in the same racial pattern as its residents?

Answer: If the answer to the question is "yes," then among the 350 new employees one would expect approximately .64(350) = 224 white, .25(350) = 87.5 black, and .11(350)= 38.5 Hispanic. Chi-square gives a measurement of the differences between these expected numbers and the actual results.

$$\chi^2 = \frac{(243 - 224)^2}{224} + \frac{(80 - 87.5)^2}{87.5} + \frac{(27 - 38.5)^2}{38.5} = 5.69$$

The smaller the resulting χ^2-value, the better the fit. To decide how large a calculated χ^2-value must be to be significant, that is, to choose a critical value, we must understand how χ^2-values are distributed. A χ^2-distribution is not symmetrical and is always skewed to the right. Just as was the case with the *t*-distribution, there are distinct χ^2-distributions each with an associated df value (number of degrees of freedom). The larger the df value, the closer the χ^2-distribution is to a normal distribution.

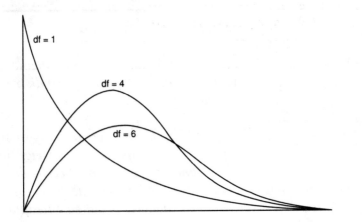

A large value of χ^2 may or may not be significant—the answer depends on which χ^2-distribution we are using. As with the t-distribution, Table C in the Appendix simply gives critical χ^2-values for the more commonly used percentages or probabilities.

There is a relationship between the χ^2-distribution and the normal distribution. While we do not give the mathematical explanation here, note that squaring our often used z-scores 1.645, 1.96, and 2.576, results in 2.706, 3.841, and 6.635, which are entries found in the first row of the χ^2-distribution table.

Key 53 Goodness-of-fit test for
uniform distribution

OVERVIEW *Are muggings evenly distributed during the week? Do partners in a law firm bring in equal numbers of new clients? Do different brands of chemical fertilizers result in similar crop yields? In many examples the interest is in whether a **uniform** distribution is present. A hypothesis test is run on a sample, and chi-square is used to judge the significance of how far the observations differ from an even distribution.*

Our approach is similar to that developed earlier. There is the null hypothesis of a good fit, that is, that a uniform distribution correctly describes the situation, problem, or activity under consideration. Our observed data consist of one possible sample from a whole universe of possible samples. We ask about the chance of obtaining a sample with our observed discrepancies if the null hypothesis is really true. Finally, if the chance is too small, we reject the null hypothesis and say that the fit is not a good one.

KEY EXAMPLE

A grocery store manager wishes to determine whether a certain product will sell equally well in any of five locations in the store. Five displays are set up, one in each location, and the resulting numbers of the product sold are noted.

Location

	1	2	3	4	5
Number sold	43	29	52	34	48

Is there enough evidence that location makes a difference? Test at both the 5% and 10% significance levels.

Answer: We note that a total of $43 + 29 + 52 + 34 + 48 = 206$ units were sold. If location doesn't matter, we would expect $206/5 = 41.2$ units sold per location (uniform distribution).

	Location				
	1	2	3	4	5
Expected number	41.2	41.2	41.2	41.2	41.2

Thus

$$\chi^2 = \frac{(43-41.2)^2}{41.2} + \frac{(29-41.2)^2}{41.2} + \frac{(52-41.2)^2}{41.2} + \frac{(34-41.2)^2}{41.2}$$

$$+ \frac{(48-41.2)^2}{41.2} = 8.903$$

The degrees of freedom are the number of classes minus 1, that is, df = 5 − 1 = 4.

H_0: good fit for uniform distribution

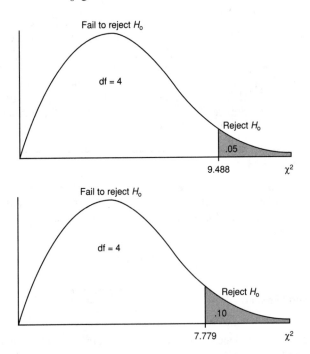

At the 5% level, with df = 4, the critical χ^2-value is 9.488, while at the 10% level the critical value is 7.779. Since 8.903 < 9.488 but 8.903 > 7.779, there is sufficient evidence to reject H_0 at the 10% level, but not at the 5% level. If the grocery store manager is willing to accept a 10% chance of committing a Type I error, then there is enough evidence to claim that location makes a difference.

Key 54 Goodness-of-fit test for

prior distribution

OVERVIEW *Have the comparative percentages of college graduates who plan to become doctors, lawyers, and corporate executives changed over the percentages of 10 years ago? Do different species of fruit flies appear in the same ratios as they did before exposure to a particular chemical? Have the percentages of voters registered as Democrat, Republican, and Independent changed since before the last presidential election? To test such hypotheses we change the prior percentages into "expected" numbers by multiplying by the sample size, and then run a chi-square analysis.*

KEY EXAMPLE

Last year, at the 6 P.M. time slot, television channels 2, 3, 4, and 5 captured the entire audience with 30%, 25%, 20%, and 25%, respectively. During the first week of the new season, 500 viewers are interviewed. If viewer preference hasn't changed, what number is expected to watch each channel?

Answer:

.30(500) = 150,   .25(500) = 125,   .20(500) = 100,   .25(500) = 125

so we have

	Channel			
	2	3	4	5
Expected number	150	125	100	125

Suppose that the actual observed numbers are as follows:

	Channel			
	2	3	4	5
Observed number	139	138	112	111

Do these numbers indicate a change? Are the differences significant? We calculate:

$$\chi^2 = \sum \frac{(\text{obs.} - \text{exp.})^2}{\text{exp.}}$$

$$= \frac{(139 - 150)^2}{150} + \frac{(138 - 125)^2}{125} + \frac{(112 - 100)^2}{100} + \frac{(111 - 125)^2}{125}$$

$$= 5.167$$

Is 5.167 large enough for us to reject the null hypothesis of a good fit between the observed and the expected? To use the χ^2-table (Table C) in the Appendix, we must decide upon an α-risk and calculate the df value. The number of degrees of freedom for this problem is $4 - 1 = 3$. [Note that, while the observed values of 139, 138, and 112 can be freely chosen, the fourth value of 111 is absolutely determined because the total must be 500; thus df = 3.]

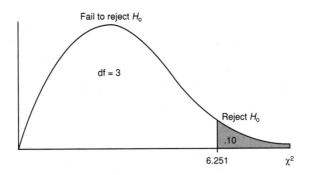

Thus, for example, at a 10% significance level, with df = 3, the critical χ^2 is 6.251. Since $5.167 < 6.251$, there is *not* sufficient evidence to reject H_0.

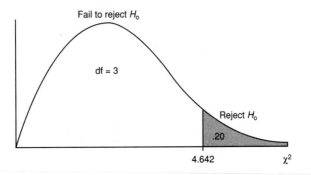

However, with df = 3 and a 20% significance level, the critical χ^2 is 4.642. Since 5.167 > 4.642, there *is* sufficient evidence to reject H_0. At the 20% level we would conclude that the fit is not good, and that viewer preference has changed.

KEY EXAMPLE

A geneticist claims that four species of fruit flies should appear in the ratio 1:3:3:9. Suppose that a sample of 4000 flies contained 226, 764, 733, and 2277 flies of each species, respectively. At the 10% significance level, is there sufficient evidence to reject the geneticist's claim?

Answer: Since $1 + 3 + 3 + 9 = 16$, according to the geneticist the expected number of fruit flies of each species is as follows:

$$\frac{1}{16}(4000) = 250, \quad \frac{3}{16}(4000) = 750, \quad \frac{3}{16}(4000) = 750,$$
$$\frac{9}{16}(4000) = 2250$$

We calculate chi-square:

$$\chi^2 = \frac{(226 - 250)^2}{250} + \frac{(764 - 750)^2}{750} + \frac{(733 - 750)^2}{750}$$
$$+ \frac{(2277 - 2250)^2}{2250} = 3.27$$

With $4 - 1 = 3$ degrees of freedom and $\alpha = .10$, the critical χ^2 value is 6.251. Since 3.27 < 6.251, there is *not* sufficient evidence to reject H_0. At the 10% significance level, the geneticist's claim should not be rejected.

Key 55 Goodness-of-fit test for standard probability distributions

OVERVIEW *Chi-square is used to test whether an observed distribution fits the binomial, the Poisson, or the normal distribution.*

KEY EXAMPLE

Suppose that a commercial is run once on television, once on the radio, and once in a newspaper. The advertising agency believes that any potential consumer has a 20% chance of seeing the ad on TV, a 20% chance of hearing it on the radio, and a 20% chance of reading it in the paper. In a telephone survey of 800 consumers, the numbers claiming to have been exposed to the ad 0, 1, 2, or 3 times is as follows:

	0	1	2	3
Observed number of people	434	329	35	2

At the 1% significance level, test the null hypothesis that the number of times any consumer saw the ad follows a binomial distribution with $\pi = 0.2$.

Answer: The complete binomial distribution with $\pi = .2$ and $n = 3$ is as follows:

$$P(0) = \quad (.8)^3 \quad = .512$$
$$P(1) = 3(.2)(.8)^2 = .384$$
$$P(2) = 3(.2)^2(.8) = .096$$
$$P(3) = \quad (.2)^3 \quad = .008$$

Multiplying each of these probabilities by 800 gives the expected number of occurrences:

$$.512(800) = 409.6, \quad .384(800) = 307.2, \quad .096(800) = 76.8,$$
$$.008(800) = 6.4$$

	0	1	2	3
Expected number of people	409.6	307.2	76.8	6.4

$$\chi^2 = \frac{(434 - 409.6)^2}{409.6} + \frac{(329 - \overline{307.2})^2}{307.2} + \frac{(35 - 76.8)^2}{76.8} + \frac{(2 - 6.4)^2}{6.4}$$

$$= 28.776$$

H_0: good fit to a binomial distribution with $\pi = .2$

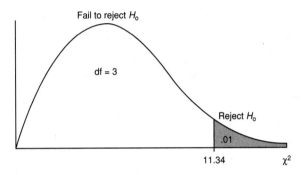

With df $= 4 - 1 = 3$ and $\alpha = .01$, the critical χ^2-value is 11.34. Since $28.776 > 11.34$, there *is* sufficient evidence to reject H_0 and to conclude that the number of ads seen by each consumer does *not* follow a binomial distribution with $\pi = .2$.

KEY EXAMPLE

For a hospital study of 450 patients with a particular form of cancer, the following chart shows the number of these patients who survived each of the given numbers of years.

	Survival Years				
	0	1	2	3	4 or more
Observed number of patients	60	110	125	88	67

Test the null hypothesis that the distribution follows a Poisson distribution with $\mu = 2.1$. Assume a 2.5% significance level.

Answer: The Poisson probabilities are:

$$P(0) = \quad e^{-2.1} \quad = .122$$

$$P(1) = \quad 2.1e^{-2.1} \quad = .257$$

$$P(2) = \frac{(2.1)^2}{2} e^{-2.1} = .270$$

$$P(3) = \frac{(2.1)^3}{3!} e^{-2.1} = .189$$

$$P(4 \text{ or more})(= 1 - (.122 + .257 + .270 + .189) = .162$$

Multiplying each of these probabilities by 450 gives the number of patients expected to survive each of the designated numbers of years:

$$.122(450) = 54.9, \quad .257(450) = 115.65, \quad .270(450) = 121.5,$$
$$.189(450) = 85.05, \quad .162(450) = 72.9.$$

Survival Years

	0	1	2	3	4 or more
Expected number of patients	54.9	115.65	121.5	85.05	72.9

Thus

$$\chi^2 = \frac{(60 - 54.9)^2}{54.9} + \frac{(110 - 115.65)^2}{115.65} + \frac{(125 - 121.5)^2}{121.5}$$

$$+ \frac{(88 - 85.05)^2}{85.05} + \frac{(67 - 72.9)^2}{72.9} = 1.430$$

H_0: good fit to a Poisson distribution with $\mu = 2.1$

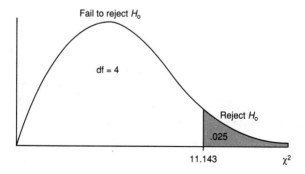

The degrees of freedom are the number of classes minus 1, that is, df = 5 − 1 = 4. With α = .025, the critical χ^2−value is 11.143. Since 1.430 < 11.143, there is *not* sufficient evidence to reject H_0. It should *not* be disputed that a Poisson distribution with μ = 2.1 describes the distribution of survival years of patients stricken with this cancer.

KEY EXAMPLE

A sample of 225 bags of rice labeled as containing 50 pounds each are weighed with the following results:

Weight (pounds)

Observed	under 49.25	49.25–49.75	49.75–50.25	50.25–50.75	over 50.75
number of bags	25	61	70	59	10

Test the null hypothesis that the distribution follows a normal distribution with μ = 50 and σ = 0.5. Assume a 2.5% and then a 1% significance level.

Answer: The z-scores of 50.25 and 50.75 are, respectively,

$$\frac{50.25 - 50}{0.5} = 0.5 \quad \text{and} \quad \frac{50.75 - 50}{0.5} = 1.5$$

Similarly, 49.75 and 49.25 have z-scores of −0.5 and −1.5. Using Table A for normal curves we find:

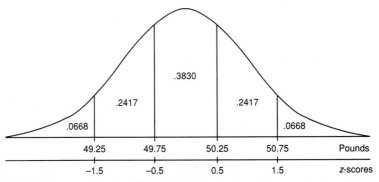

Multiplying each probability by 225 bags gives the expected numbers of bags for the corresponding weight ranges:

$$.0668(225) = 15.03, \quad .2417(225) = 54.38, \quad .3830(225) = 86.18$$

Weight (pounds)

	under 49.25	49.25–49.75	49.75–50.25	50.25–50.75	over 50.75
Expected number of bags	15.03	54.38	86.18	34.38	15.03

Thus

$$\chi^2 = \frac{(25-15.03)^2}{15.03} + \frac{(61-54.38)^2}{54.38} + \frac{(70-86.18)^2}{86.18}$$

$$+ \frac{(59-54.38)^2}{54.38} + \frac{(10-15.03)^2}{15.03} = 12.533$$

H_0: good fit to a normal distribution with $\mu = 50$ and $\sigma = 0.5$

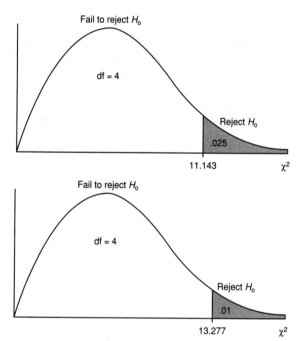

With df = 5 − 1 = 4 and $\alpha = .025$, the critical χ^2–value is 11.143, while $\alpha = .01$ results in a critical value of 13.277. Since 12.533 > 11.143 and 12.533 < 13.277, we have sufficient evidence to reject H_0 at the 2.5% significance level but not at the 1% level. Thus, if we are willing to accept a 2.5% α-risk, we should conclude that the distribution of weights of rice bags is *not* normal.

Key 56   Fits that are too good

OVERVIEW   *We have been using only small, upper-tail probabilities of the chi-square distribution. There are cases, however, where we are interested in the probability that a chi-square value is less than or equal to a critical value.*

KEY EXAMPLE

Suppose your mathematics instructor gives, as a homework assignment, the problem of testing the fairness of a certain die. You are asked to roll the die 6000 times and note how often it comes up 1, 2, 3, 4, 5, and 6. The assignment tires you quickly, so you decide to simply invent some "reasonable" data. Being careful to make the total 6000, you write:

Number of times observed	1	2	3	4	5	6
	988	991	1010	990	1013	1008

H_0: good fit with uniform distribution, that is, the die is fair

The "expected" values are all 1000, and so

$$\chi^2 = \frac{12^2}{1000} + \frac{9^2}{1000} + \frac{10^2}{1000} + \frac{10^2}{1000} + \frac{13^2}{1000} + \frac{8^2}{1000} = 0.658$$

What is the conclusion if $\alpha = .01$? $.05$? $.10$?

Answer: With df = 6 − 1 = 5, the critical χ^2-values are 15.086, 11.070, and 9.236. Since 0.658 is less than each of these, there is not enough evidence to claim that the die is unfair.

What if we're willing to accept $\alpha = .90$? $.95$? $.975$? $.99$?

Answer: These give critical chi-square values of 1.610, 1.145, 0.831, and 0.554, so only if we accept a 99% chance of committing a Type I error can we reject H_0 and question the fairness of the die. Even if the die is fair, the probability of obtaining such a "good-fitting" sample is extremely small. Your instructor would reasonably conclude that you never completed the assignment, but simply made up the data. Your results are too good to be true!

Key 57 Independence and contingency tables

OVERVIEW *In each of the goodness-of-fit problems of Keys 53–56, there was a set of expectations based on some assumption about how the distribution should turn out. We then tested whether or not an observed sample distribution might reasonably have come from a larger set based on the assumed distribution. In many real-world problems, however, we want simply to compare two or more observed samples without any prior assumptions about an expected distribution. In what is called a **test of independence**, we ask whether the two or more samples might have reasonably come from some one larger set. For example, do students, professors, and administrators all have the same opinion concerning the need for a new science building? Do nonsmokers, light smokers, and heavy smokers all have the same likelihood of being eventually diagnosed with cancer, heart disease, or emphysema?*

We classify our observations in two ways, and then ask whether the two ways are independent. For example, we might consider several age groups, and within each age group ask how many employees show various levels of job satisfaction. The null hypothesis would be that age and job satisfaction are independent, that is, that the proportion of employees expressing a given level of job satisfaction is the same no matter which age group is considered. A sociologist might classify people by ethnic origin, and within each group ask how many individuals complete various levels of education. The null hypothesis would be that ethnic origin and education level are independent, that is, that the proportion of people achieving a given education level is the same no matter which group is considered.

Our analysis will involve calculating a table of *expected* values, assuming the null hypothesis about independence is true. We compare these expected values with the observed values, and ask whether the differences are reasonable if H_0 is true. The significance of the differences is gauged by the same χ^2-value of weighted squared differences used in the preceding keys. The smaller the resulting χ^2-value, the more reasonable is the null hypothesis of independence. If the χ^2-value exceeds some critical value, then we can say that there is sufficient evidence to reject the null hypothesis and claim that there *is* some relationship between the two variables or methods of classification.

KEY EXAMPLE

A beef distributor wishes to determine whether there is a relationship between geographic region and cut of meat preferred. If there is no relationship, we will say that beef preference is *independent* of geographic region. Suppose that, in a random sample of 500 consumers, 300 are from the North while 200 are from the South. Of these, 150 prefer cut A, 275 prefer cut B, and only 75 prefer cut C.

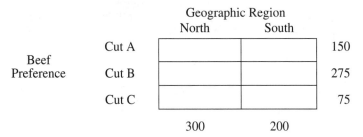

| | Geographic Region | | |
	North	South	
Cut A			150
Cut B			275
Cut C			75
	300	200	

Beef Preference

If beef preference is independent of geographic region, how would we expect this table to be filled in?

Answer: Since 300/500 = .6 of the sample is from the North, we would expect .6 of the 150 consumers favoring cut A to be from the North. We calculate .6(150) = 90 or (300)(150)/500 = 90. Similarly, we would expect .6 of the 275 consumers favoring cut B to be from the North. We calculate .6(275) = 165 or (300)(275)/500 = 165. Continuing in this manner we fill in the table as follows:

Expected results:

	North	South	
Cut A	90	60	150
Cut B	165	110	275
Cut C	45	30	75
	300	200	

Note that we didn't have to actually perform all the calculations to fill in the table. After determining 90 and 165, there were really no choices left for the remaining values because the row totals and column totals were already set. Thus, in this problem there are *two* degrees of freedom, that is, df = 2.

Now suppose that in the actual sample of 500 consumers the observed numbers were as follows:

Observed results:

	North	South	
Cut A	100	50	150
Cut B	150	125	275
Cut C	50	25	75
	300	200	

Are the differences between the expected and observed values large or small? If the differences are large enough, we will reject independence and claim that beef preference is related to geographic location. We calculate the χ^2-value:

$$\chi^2 = \sum \frac{(\text{obs.} - \text{exp.})^2}{\text{exp.}}$$

$$= \frac{(100-90)^2}{90} + \frac{(50-60)^2}{60} + \frac{(150-165)^2}{165} + \frac{(125-110)^2}{110}$$

$$+ \frac{(50-45)^2}{45} + \frac{(25-30)^2}{30}$$

$$= 7.578$$

Is 7.578 large enough for us to reject the null hypothesis of independence?

Answer:

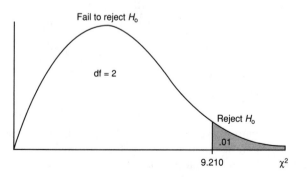

Looking at the χ^2 table, with df = 2, we see that, with $\alpha = .01$, the critical χ^2 is 9.210. Thus there is not enough evidence to reject H_0.

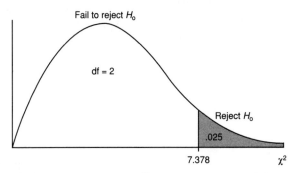

However, with $\alpha = .025$, the critical χ^2 is 7.378; and since 7.578 > 7.378, there is sufficient evidence to reject H_0. Thus, we can conclude that at the 2.5% significance level there is a relationship between beef preference and geographic region, while at the 1% significance level there is *not* sufficient evidence to reject the null hypothesis of independence.

KEY EXAMPLE

In a nationwide telephone poll of 1000 adults, representing Democrats, Republicans, and Independents, respondents were asked if their confidence in the U.S. banking system had been shaken by the Savings and Loan crisis. The answers, cross-classified by party affiliation, are given in the following *contingency* table.

Observed results:

	Yes	No	No opinion
Democrats	175	220	55
Republicans	150	165	35
Independents	75	105	20

Test the null hypothesis that shaken confidence in the banking system is independent of party affiliation. Use a 10% significance level.

Answer: The above table gives the *observed* results. To find the *expected* values, we must first determine the row and column totals, which were given in the preceding example.

We obtain the following values:

Row totals: 175 + 220 + 55 = 450, 150 + 165 + 35 = 350,
                75 + 105 + 20 = 200

Column totals: 175 + 150 + 75 = 400, 220 + 165 + 105 = 490,
                55 + 35 + 20 = 110.

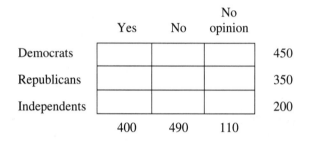

	Yes	No	No opinion	
Democrats				450
Republicans				350
Independents				200
	400	490	110	

To calculate, for example, the expected value in the upper left box, we can proceed in any of several equivalent ways. First, we could note that the proportion of Democrats is 450/1000 = .45; and so, if independent, the expected number of Democrat "yes" responses is .45(400) = 180. Instead, we could note that the proportion of "yes" responses is 400/1000 = .4; and so, if independent, the expected number of Democrat "yes" responses is .4(450) = 180. Finally, we could note that both these calculations simply involve (450)(400)/1000 = 180. In other words, the expected value of any box can be calculated by multiplying its corresponding row total times the appropriate column total and then dividing by the grand total. Thus, for example, the expected value for the middle box corresponding to Republican "no" responses is (350)(490)/1000 = 171.5. Continuing in this manner, we find:

Expected results:

	Yes	No	No opinion	
Democrats	180	220.5	49.5	450
Republicans	140	171.5	38.5	350
Independents	80	98	22	200
	400	490	110	

Then
$$\chi^2 = \frac{(175-180)^2}{180} + \frac{(220-220.5)^2}{220.5} + \frac{(55-49.5)^2}{49.5} + \frac{(150-140)^2}{140}$$
$$+ \frac{(165-171.5)^2}{171.5} + \frac{(35-38.5)^2}{38.5} + \frac{(75-80)^2}{80} + \frac{(105-98)^2}{98}$$
$$+ \frac{(20-22)^2}{22}$$
$$= 3.024$$

Note that, when the 180, 220.5, 140, and 171.5 boxes were calculated, the other expected values could be found by using the row and column totals. Thus, the number of degrees of freedom here is 4. More generally, in this type of problem

$$df = (r-1)(c-1)$$

where r is the number of rows and c is the number of columns.

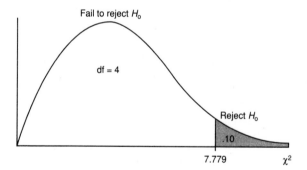

With $\alpha = .10$ and df = 4, the critical χ^2-value is 7.779. Since 3.024 < 7.779, there is *not* sufficient evidence to reject the null hypothesis of independence. Thus, at the 10% significance level, we *cannot* claim a relationship between party affiliation and shaken confidence in the banking system.

Two points are worth noting:

- If any of the *expected* values are too small, the χ^2-value tends to turn out to be unfairly high. The usual rule of thumb cut-off point is taken to be 5; that is, this procedure is not used if any expected value is below 5. Sometimes this difficulty is overcome by a regrouping that combines two or more classifications. However, this also leads to smaller numbers of degrees of freedom and less information from the tables.

- Even when we have sufficient evidence to reject the null hypothesis of independence, we cannot necessarily claim any direct *causal* relationship. In other words, while we do make a statement about some link or relationship between two variables, we are not justified in claiming that one is causing the other. For example, we may show a relationship between salary level and job satisfaction, but our methods would not show that higher salaries *cause* higher job satisfaction. Perhaps if an employee has higher job satisfaction, this causes his superiors to be impressed and thus leads to larger increases in pay. Or perhaps there is a third variable, such as training, education, or personality, that has a direct causal relationship with both salary level and job satisfaction.

Key 58   Theme exercises with answers

OVERVIEW   *Sample questions of the type that might appear on homework assignments and tests are presented with answers.*

- A pet food manufacturer runs an experiment to determine whether three brands of dog food are equally preferred. In the experiment, 150 dogs are individually set loose in front of three dishes of food and their choices are noted. Tabulations show that 62 dogs went to brand A, 43 to brand B, and 45 to brand C. Is there sufficient evidence to say that dogs have preferences among the brands? Test at the 2.51% significance level.

  Answer: If dogs have no preferences among the three brands, we would expect 150/3 = 50 dogs to go to each dish. Thus:

  $$\chi^2 = \frac{(62-50)^2}{50} + \frac{(43-50)^2}{50} + \frac{(45-50)^2}{50} = 4.36$$

  With 3 − 1 = 2 degrees of freedom and $\alpha = .025$, the critical χ^2-value is 7.378. Since 4.36 < 7.378, there is *not* sufficient evidence to say that dogs have preferences among the three given brands. [Of course the company manufacturing brand A will continue to claim that more dogs prefer its food!]

- A highway superintendent states that five bridges into a city are used in the ratio 2:3:3:4:6 during the morning rush hour. A highway study of a sample of 9000 cars indicates that 1070, 1570, 1513, 1980, and 2867 cars, respectively, use the five bridges. Can the superintendent's claim be rejected if $\alpha = .05$? .025?

  Answer: Since 2 + 3 + 3 + 4 + 6 = 18, according to the superintendent the expected number of cars using each bridge is as follows:

  $$\frac{2}{18}(9000) = 1000, \quad \frac{3}{18}(9000) = 1500, \quad 1500,$$

  $$\frac{4}{18}(9000) = 2000, \quad \frac{6}{18}(9000) = 3000$$

We calculate chi-square:

$$\chi^2 = \frac{(1070-1000)^2}{1000} + \frac{(1570-1500)^2}{1500} + \frac{(1513-1500)^2}{1500}$$

$$+ \frac{(1980-2000)^2}{2000} + \frac{(2867-3000)^2}{3000}$$

$$= 14.38$$

With $5 - 1 = 4$ degrees of freedom and $\alpha = .01$, the critical χ^2-value is 13.277. Since $14.38 > 13.277$, there *is* sufficient evidence to reject H_0 and to claim that the superintendent is wrong.

- Four commercial flights per day are made from a small county airport. Suppose the airport manager tabulates the number of on-time departures each day for 200 days.

Number of On-Time Departures

	0	1	2	3	4
Observed number of days	13	36	72	56	23

At the 5% significance level, test the null hypothesis that the daily distribution is binomial.

Answer: In this problem, we are not given a value of π to test against, so we use the observed sample to find a proportion $\bar{p}$. The number of on-time departures is seen to be $13(0) + 36(1) + 72(2) + 56(3) + 23(4) = 440$. Since there were $4(200) = 800$ flights, this gives a proportion of

$$\bar{p} = \frac{440}{800} = .55$$

The binomial distribution with $\pi = .55$ and $n = 4$ is

$$P(0) = (.45)^4 \qquad = .0410$$
$$P(1) = 4(.55)(.45)^3 \quad = .2005$$
$$P(2) = 6(.55)^2(.45)^2 = .3675$$
$$P(3) = 4(.55)^3(.45) \quad = .2995$$
$$P(4) = (.55)^4 \qquad = .0915$$

Multiplying each of these probabilities by 200 days gives the expected numbers of occurrences: .0410(200) = 8.2, .2005(200) = 40.1, .3675(200) = 73.5, .2995(200) = 59.9, .0915(200) = 18.3.

Number of On-Time Departures

	0	1	2	3	4
Expected number of days	8.2	40.1	73.5	59.9	18.3

We calculate chi-square:

$$\chi^2 = \frac{(13 - 8.2)^2}{8.2} + \frac{(36 - 40.1)^2}{40.1} + \frac{(72 - 73.5)^2}{73.5} + \frac{(56 - 59.9)^2}{59.9}$$

$$+ \frac{(23 - 18.3)^2}{18.3} = 4.721$$

H_0: good fit with a binomial

In this case the sample not only gives the total number, 200, but also gives the proportion, .55. Thus the number of degrees of freedom is the number of classes minus 2, that is, df = 5 − 2 = 3. With $\alpha = .05$, this gives a critical χ^2-value of 7.815. Since 4.721 < 7.815, there is *not* sufficient evidence to reject H_0. The observed data do *not* differ significantly from what would be expected for a binomial distribution.

• An editor checks the number of typing errors per page on 125 pages of a manuscript.

Number of errors

	0	1	2	3	4	5
Observed number of pages	29	38	24	18	13	3

Does the number of errors follow a Poisson distribution? Test at the 10% significance level.

Answer: In this case the mean is not given, so it must be estimated from the sample. The total number of errors is 29(0) + 38(1) + 24(2) + 18(3) + 10(4) + 3(5) = 195. Thus the sample average per page is 195/125 = 1.56. The Poisson probabilities are:

$$P(0) = e^{-1.56} = .210$$

$$P(1) = 1.56e^{-1.56} = .328$$

$$P(2) = \frac{(1.56)^2}{2}e^{-1.56} = .256$$

$$P(3) = \frac{(1.56)^3}{3!}e^{-1.56} = .133$$

$$P(4) = \frac{(1.56)^4}{4!}e^{-1.56} = .052$$

$P(5 \text{ or more}) = 1 - (.210 + .328 + .256 + .133 + .052) = .021$

Multiplying by 125 pages gives the expected number of pages with various numbers of mistakes: $.210(125) = 26.25$, $.328(125) = 41$, $.256(125) = 32$, $.133(125) = 16.625$, $.052(125) = 6.5$, and $.021(125) = 2.625$.*

Number of errors

	0	1	2	3	4	5
Expected number of pages	26.25	41	32	16.625	6.5	2.625

Thus

$$\chi^2 = \frac{(29-26.25)^2}{26.25} + \frac{(38-41)^2}{41} + \frac{(24-32)^2}{32} + \frac{(18-16.625)^2}{16.625}$$
$$+ \frac{(13-6.5)^2}{6.5} + \frac{(3-2.625)^2}{2.625}$$
$$= 9.175$$

H_0: good fit with a Poisson distribution

Since the sample is giving both the total number of pages (125) and the average errors per page (1.56), the number of degrees of freedom is the number of classes minus 2, that is, df = 6 − 2 = 4. With $\alpha = .10$, the critical χ^2-value is 7.779. Since $9.175 > 7.779$, there

---

*Note that this value violates the previously mentioned rule of thumb that every cell must have an expected value of at least 5. Some statisticians do accept one lower value; alternatively, we could combine the last two cells into one with a higher expected value.

is sufficient evidence at the 5% level to reject H_0 and to claim that the typing errors per page do *not* follow the Poisson distribution.

- Suppose that the assembly times for a sample of 300 units of an electronic product have mean $\mu = 84$, standard deviation $\sigma = 3$, and the following distribution:

Assembly Time (minutes)

	< 78	78–81	81–84	84–87	87–90	>90
Observed number of units	15	39	87	96	48	15

At the 1% significance level, test the null hypothesis that the distribution is normal.

Answer: Here we must use the sample mean, 84, and the sample standard deviation, 3. The normal probability table gives:

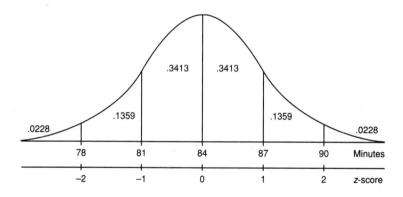

Multiplying by 300 units yields the following expected numbers of units.

Assembly Time (minutes)

	< 78	78–81	81–84	84–87	87–90	>90
Expected number of units	6.84	40.77	102.39	102.39	40.7	6.84

Thus

$$\chi^2 = \frac{(15-6.84)^2}{6.84} + \frac{(39-40.77)^2}{40.77} + \frac{(87-102.39)^2}{102.39} + \frac{(96-102.39)^2}{102.39}$$

$$+ \frac{(48-40.77)^2}{40.77} + \frac{(15-6.84)^2}{6.84}$$

$$= 23.540$$

Since we are using three measures from the sample (size, mean, and standard deviation), the number of degrees of freedom equals the number of classes minus 3, that is, df = 6 − 3 = 3. With α = .01, the critical χ^2-value is 11.34. Since 23.540 > 11.34, there *is* sufficient evidence to reject H_0 and to conclude that the distribution of assembly times is *not* normal.

• A medical researcher tests 640 heart-attack victims for the presence of a certain antibody in their blood and cross-classifies against the severity of the attack. The results are reported in the following table:

Observed results: Severity of Attack

		Severe	Medium	Mild
	Positive	85	125	150
Antibody Test				
	Negative	40	95	145

Is there evidence of a relationship between presence of the antibody and severity of the heart attack? Test at the 5% significance level.

Answer:

H_0: independence (no relation between antibody and attack)

H_a: dependence (attack severity related to presence of antibody)

$$\alpha = .05, \quad df = (2-1)(3-1) = 2$$

Totaling rows and columns yields:

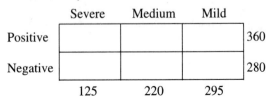

	Severe	Medium	Mild	
Positive				360
Negative				280
	125	220	295	

The expected value for each box is calculated by multiplying row total by column total and dividing by 640:

Expected results:

	Severe	Medium	Mild	
Positive	70.3	123.8	165.9	360
Negative	54.7	96.2	129.1	280
	125	220	295	

Then

$$\chi^2 = \frac{(85-70.3)^2}{70.3} + \frac{(125-123.8)^2}{123.8} + \frac{(150-165.9)^2}{165.9}$$
$$+ \frac{(40-54.7)^2}{54.7} + \frac{(95-96.2)^2}{96.2} + \frac{(145-129.1)^2}{129.1}$$
$$= 10.533$$

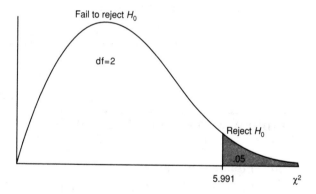

With $\alpha = .05$ and df = 2, the critical χ^2-value is 5.991. Since $10.533 > 5.991$, there is sufficient evidence to reject the null hypothesis of independence. Thus, at the 5% significance level, there *is* a relationship between presence of the antibody and severity of the heart attack.

Theme 9 REGRESSION ANALYSIS

*M*any decisions are based on a perceived relationship between two variables. For example, a company's market share may vary directly with advertising expenditures. A person's blood pressure may increase or decrease inversely as less or more hypertension medication is taken. A student's grades will probably go up and down according to the number of hours of studying time per week.

Two questions arise. First, how can the strength of an apparent relationship be measured? Second, how can an observed relationship be put into functional terms? For example, not only may a real estate broker wish to determine whether a relationship exists between the prime rate and the number of new homes sold in a month, but also it is useful to develop an expression with which to predict the number of house sales given a particular value of the prime rate.

Key 59   Scatter diagrams

OVERVIEW *Suppose a relationship is perceived between two variables called X and Y, and we graph the pairs (x,y). The result, called a **scatter diagram**, gives a visual impression of the existing relationship between the variables.*

KEY EXAMPLE

The following table gives the ages and salaries (in $1000's) of four executives in a business firm.

Age	38	53	42	47
Salary	45	86	58	61

Plotting the four points (38,45), (53,86), (42,58), and (47,61) gives the following *scatter diagram:*

In the following keys, we will be interested in finding the *best-fitting* straight line that can be drawn through such a scatter diagram.

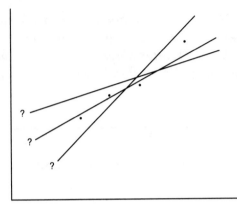

Key 60   Equation of the regression line

OVERVIEW   *The **best-fitting** straight line, that is, the line that minimizes the sum of the squares of the differences between the observed values and the values predicted by the line, is called the **regression line**.*

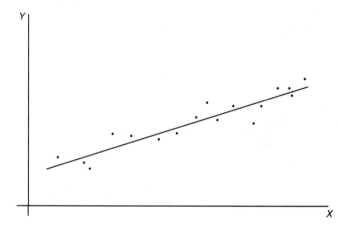

It is reasonable, intuitive, and correct that the best-fitting line will pass through $(\bar{x}, \bar{y})$, where $\bar{x}$ and $\bar{y}$ are the means of the variables X and Y. Then, from the basic expression for a line with a given slope through a given point, comes the equation

$$y' - \bar{y} = m(x - \bar{x})$$

or

$$y' = \bar{y} + m(x - \bar{x})$$

where m is the slope of the line.

It can be shown algebraically that

$$m = \frac{\sum xy - n\,\bar{x}\,\bar{y}}{\sum x^2 - n(\bar{x})^2}$$

where

$$\sum xy = x_1 y_1 + x_2 y_2 + \cdots + x_n y_n \quad \text{and} \quad \sum x^2 = x_1^2 + x_2^2 + \cdots + x_n^2$$

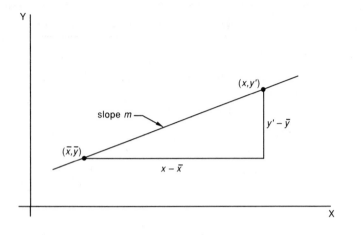

KEY EXAMPLE

An insurance company conducts a survey of 15 of its life insurance agents. The average number of minutes spent with each potential customer and the number of policies sold in a week are noted for each agent. Letting *X* and *Y* represent the average number of minutes and the number of sales, respectively, we have:

X	25	23	30	25	20	33	18	21	22	30	26	26	27	29	20
Y	10	11	14	12	8	18	9	10	10	15	11	15	12	14	11

Find the equation of the best-fitting straight line for this data.

Answer: Plotting the 15 points, (25,10), (23,11), . . . , (20,11), gives an intuitive visual impression of the relationship:

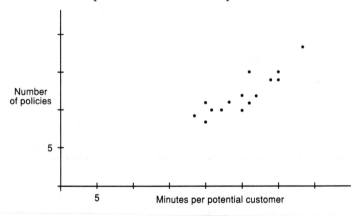

184

The above scatter diagram indicates the existence of a relationship that appears to be *linear*; that is, the points lie roughly on a straight line. Furthermore, the linear relationship is *positive*; that is, as one variable increases, so does the other (the straight line slopes upward).

To determine the equation of the best-fitting line, we first calculate:

$$\sum x = 25 + 23 + 30 + \cdots + 20 = 375$$
$$\sum y = 10 + 11 + 14 + \cdots + 11 = 180$$
$$\sum x^2 = 25^2 + 23^2 + 30^2 + \cdots + 20^2 = 9639$$
$$\sum xy = 25(10) + 23(11) + 30(14) + \cdots + 20(11) = 4645$$

Then $n = 15$, so

$$\bar{x} = \frac{\sum x}{n} = \frac{375}{15} = 25, \quad \bar{y} = \frac{\sum y}{n} = \frac{180}{15} = 12$$

and

$$m = \frac{\sum xy - n\bar{x}\bar{y}}{\sum x^2 - n(\bar{x})^2} = \frac{4645 - 15(25)(12)}{9639 - 15(25)^2} = \frac{145}{264} = 0.5492$$

Thus the regression line is given by

$$y' = \bar{y} + m(x - \bar{x}) = 12 + 0.5492(x - 25) = 0.5492x - 1.73$$

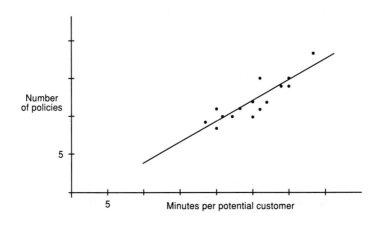

Key 61 Slope of and predictions from the regression line

OVERVIEW *The regression line enables us to predict* Y *values from given* X *values. The slope of the regression line yields a numerical insight into the relationship between the variables.*

KEY EXAMPLE

In the example of Key 60, the regression line was calculated to be $y' = 0.5492x - 1.73$. We might predict that agents who average 24 minutes per customer will average $0.5492(24) - 1.73 = 11.45$ sales per week. We also note that each additional minute spent seems to produce an average 0.5492 number of extra sales.

KEY EXAMPLE

Following are advertising expenditures and total sales with regard to six detergent products.

Advertising ($1000's):	x	2.3	5.7	4.8	7.3	5.9	6.2
Total sales ($1000's):	y	77	105	96	118	102	95

Find the equation of the regression line. Interpret the slope.

Answer: We calculate

$$\Sigma x = 32.2, \quad \Sigma y = 593, \quad \Sigma xy = 3288.6, \quad \Sigma x^2 = 187.36$$

which give

$$\bar{x} = 5.367, \quad \bar{y} = 98.833, \quad m = 7.293$$

Thus

$$y' = 98.833 + 7.293(x - 5.367) = 7.293x + 56.691$$

Using the equation of the regression line, we can predict, for example, that if $5000 is spent on advertising, the resulting total sales will be $7.293(5) + 56.691 = 93.156$ thousand or $93,156.

The slope of the regression line indicates that every extra $1000 spent on advertising will result in $7293 in added sales.

Key 62   Correlation coefficient

OVERVIEW   *There are ways to gauge whether or not the relationship between variables is strong enough so that finding the regression line and making use of it are meaningful.*

One measure of an apparent relationship is called the *correlation coefficient* and is denoted as r. The value r^2 is actually the ratio of the variance of the predicted values, y', to the variance of the observed values, y.

It can be shown algebraically that

$$r^2 = \frac{\left(\sum xy - n\,\bar{x}\,\bar{y}\right)^2}{\left[\sum x^2 - n(\bar{x})^2\right]\left[\sum y^2 - n(\bar{y})^2\right]}$$

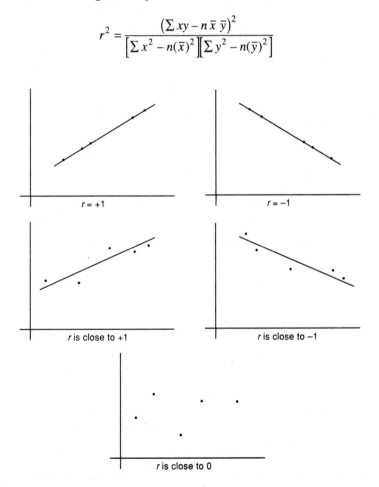

The correlation coefficient r will take the same sign as m (the slope of the regression line). The value of r always falls between -1 and $+1$, with -1 indicating perfect negative correlation, and $+1$ indicating perfect positive correlation.

KEY EXAMPLE

For the example in Key 60, we noted that $n = 15$, $\sum x = 375$, $\bar{x} = 25$, $\sum y = 180$, $\bar{y} = 12$, $\sum x^2 = 9639$, and $\sum xy = 4645$. In addition, we now calculate

$$\sum y^2 = 10^2 + 11^2 + 14^2 + \cdots + 11^2 = 2262$$

and thus

$$r^2 = \frac{(4645 - 15(25)(12))^2}{[9639 - 15(25)^2][2262 - 15(12)^2]} = \frac{(145)^2}{(264)(102)} = .7808$$

and $r = .8836$.

Key 63 Hypothesis test for correlation

OVERVIEW *To gauge the significance of the correlation coefficient, we may conduct a hypothesis test involving appropriate levels of significance.*

Is $r = .9636$ with four data pairs significant? How about $.8836$ with 15 data pairs? If there really is no correlation, what are the chances that the r-values could have been this large? These answers depend on what level of significance we are working with. In the Appendix, Table D gives critical r-values for both the 5% and 1% levels of significance; that is, with no correlation, there are $.05$ and $.01$ probabilities of obtaining the given critical levels. One also talks about α-risks of $.05$ and $.01$, respectively. Note that one must know the number of degrees of freedom, which in this situation is $n - 2$.

We have

$$H_0: \text{no correlation}$$

$$H_a: \text{correlation}$$

If our calculated r, in absolute value, is greater than the critical r, then we reject H_0 and say that at the given significance level there is sufficient evidence that r, the correlation coefficient, is significant.

KEY EXAMPLE

For the example carried through Keys 60–62, df = $15 - 2 = 13$, which gives critical values of $.514$ and $.641$ at the 5% and 1% significance levels, respectively. Both $.8836 > .514$ and $.8836 > .641$, so at either of these significance levels we conclude there *is* sufficient evidence to indicate correlation.

Key 64   Theme exercises with answers

OVERVIEW   *Sample questions of the type that might appear on homework assignments and tests are presented with answers.*

Following are the lengths and grades of ten research papers written for a sociology professor's class.

Length (pages): x	25	32	20	28	15	34	29	30	45	35
Grade: y	69	81	72	75	64	89	84	73	92	86

- Find the equation of the regression line.

- Plot a scatter diagram and graph the regression line.

- Use the equation to predict the grade for a student who turns in a paper 40 pages long.

- What is the slope of the regression line, and what does it signify?

- Test for correlation at both the 5% and 1% levels of significance.

Answer: We calculate as follows:

$$\Sigma x = 25 + 32 + \cdots + 35 = 293$$

$$\Sigma y = 69 + 81 + \cdots + 86 = 785$$

$$\Sigma xy = 25(69) + 32(81) + \cdots + 35(86) = 23{,}619$$

$$\Sigma x^2 = 25^2 + 32^2 + \cdots + 35^2 = 9{,}205$$

$$\Sigma y^2 = 69^2 + 81^2 + \cdots + 86^2 = 62{,}393$$

$$\bar{x} = \frac{293}{10} = 29.3$$

$$\bar{y} = \frac{785}{10} = 78.5$$

$$m = \frac{23,619 - 10(29.3)(78.5)}{9205 - 10(29.3)^2} = \frac{618.5}{620.1} = 0.997$$

$$y' = 78.5 + 0.997(x - 29.3) = 0.997x + 49.3$$

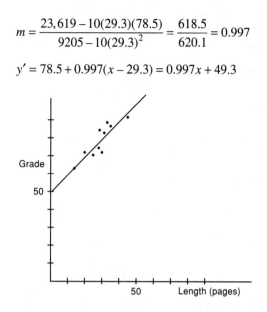

The grade for a student who turns in a paper 40 pages long is predicted to be 0.997(40) + 49.3 = 89.2.

The slope of the regression line indicates that, for each additional page, students can increase their grades by 0.997, that is, approximately a point per page.

$$r^2 = \frac{(618.5)^2}{(620.1)(62,393 - 10(78.5)^2)} = .801 \quad \text{and} \quad r = .895$$

The df = 10 − 2 = 8, which gives critical values of .632 and .765 at the 5% and 1% significance levels, respectively. Both .895 > .632 and .895 > .765, so at either of these significance levels we conclude that there *is* sufficient evidence to indicate correlation between a student's grade and the number of pages in the research paper.

APPENDIX

Table A

Normal Curve Areas

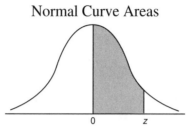

z	.00	.01	.02	.03	.04	.05	.06	.07	.08	.09
0.0	.0000	.0040	.0080	.0120	.0160	.0199	.0239	.0279	.0319	.0359
0.1	.0398	.0438	.0478	.0517	.0557	.0596	.0636	.0675	.0714	.0753
0.2	.0793	.0832	.0871	.0910	.0948	.0987	.1026	.1064	.1103	.1141
0.3	.1179	.1217	.1255	.1293	.1331	.1368	.1406	.1443	.1480	.1517
0.4	.1554	.1591	.1628	.1664	.1700	.1736	.1772	.1808	.1844	.1879
0.5	.1915	.1950	.1985	.2019	.2054	.2088	.2123	.2157	.2190	.2224
0.6	.2257	.2291	.2324	.2357	.2389	.2422	.2454	.2486	.2517	.2549
0.7	.2580	.2611	.2642	.2673	.2704	.2734	.2764	.2794	.2823	.2852
0.8	.2881	.2910	.2939	.2967	.2995	.3023	.3051	.3078	.3106	.3133
0.9	.3159	.3186	.3212	.3238	.3264	.3289	.3315	.3340	.3365	.3389
1.0	.3413	.3438	.3461	.3485	.3508	.3531	.3554	.3577	.3599	.3621
1.1	.3643	.3665	.3686	.3708	.3729	.3749	.3770	.3790	.3810	.3830
1.2	.3849	.3869	.3888	.3907	.3925	.3944	.3962	.3980	.3997	.4015
1.3	.4032	.4049	.4066	.4082	.4099	.4115	.4131	.4147	.4162	.4177
1.4	.4192	.4207	.4222	.4236	.4251	.4265	.4279	.4292	.4306	.4319
1.5	.4332	.4345	.4357	.4370	.4382	.4394	.4406	.4418	.4429	.4441
1.6	.4452	.4463	.4474	.4484	.4495	.4505	.4515	.4525	.4535	.4545
1.7	.4554	.4564	.4573	.4582	.4591	.4599	.4608	.4616	.4625	.4633
1.8	.4641	.4649	.4656	.4664	.4671	.4678	.4686	.4693	.4699	.4706
1.9	.4713	.4719	.4726	.4732	.4738	.4744	.4750	.4756	.4761	.4767
2.0	.4772	.4778	.4783	.4788	.4793	.4798	.4803	.4808	.4812	.4817
2.1	.4821	.4826	.4830	.4834	.4838	.4842	.4846	.4850	.4854	.4857
2.2	.4861	.4864	.4868	.4871	.4875	.4878	.4881	.4884	.4887	.4890
2.3	.4893	.4896	.4898	.4901	.4904	.4906	.4909	.4911	.4913	.4916
2.4	.4918	.4920	.4922	.4925	.4927	.4929	.4931	.4932	.4934	.4936
2.5	.4938	.4940	.4941	.4943	.4945	.4946	.4948	.4949	.4951	.4952
2.6	.4953	.4955	.4956	.4957	.4959	.4960	.4961	.4962	.4963	.4964
2.7	.4965	.4966	.4967	.4968	.4969	.4970	.4971	.4972	.4973	.4974
2.8	.4974	.4975	.4976	.4977	.4977	.4978	.4979	.4979	.4980	.4981
2.9	.4981	.4982	.4982	.4983	.4984	.4984	.4985	.4985	.4986	.4986
3.0	.4987	.4987	.4987	.4988	.4988	.4989	.4989	.4989	.4990	.4990

Abridged from Table I of A. Hald, *Statistical Tables and Formulas* (New York: John Wiley & Sons, Inc.), 1952. Reproduced by permission of A. Hald and the publisher, John Wiley & Sons, Inc.

Table B
Critical Values of t

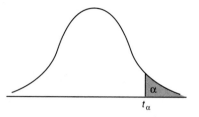

df	$t_{.100}$	$t_{.050}$	$t_{.025}$	$t_{.010}$	$t_{.005}$	$t_{.001}$	$t_{.0005}$
1	3.078	6.314	12.706	31.821	63.657	318.31	636.62
2	1.886	2.920	4.303	6.965	9.925	22.326	31.598
3	1.638	2.353	3.182	4.541	5.841	10.213	12.924
4	1.533	2.132	2.776	3.747	4.604	7.173	8.610
5	1.476	2.015	2.571	3.365	4.032	5.893	6.869
6	1.440	1.943	2.447	3.143	3.707	5.208	5.959
7	1.415	1.895	2.365	2.998	3.499	4.785	5.408
8	1.397	1.860	2.306	2.896	3.355	4.501	5.041
9	1.383	1.833	2.262	2.821	3.250	4.297	4.781
10	1.372	1.812	2.228	2.764	3.169	4.144	4.587
11	1.363	1.796	2.201	2.718	3.106	4.025	4.437
12	1.356	1.782	2.179	2.681	3.055	3.930	4.318
13	1.350	1.771	2.160	2.650	3.012	3.852	4.221
14	1.345	1.761	2.145	2.624	2.977	3.787	4.140
15	1.341	1.753	2.131	2.602	2.947	3.733	4.073
16	1.337	1.746	2.120	2.583	2.921	3.686	4.015
17	1.333	1.740	2.110	2.567	2.898	3.646	3.965
18	1.330	1.734	2.101	2.552	2.878	3.610	3.922
19	1.328	1.729	2.093	2.539	2.861	3.579	3.883
20	1.325	1.725	2.086	2.528	2.845	3.552	3.850
21	1.323	1.721	2.080	2.518	2.831	3.527	3.819
22	1.321	1.717	2.074	2.508	2.819	3.505	3.792
23	1.319	1.714	2.069	2.500	2.807	3.485	3.767
24	1.318	1.711	2.064	2.492	2.797	3.467	3.745
25	1.316	1.708	2.060	2.485	2.787	3.450	3.725
26	1.315	1.706	2.056	2.479	2.779	3.435	3.707
27	1.314	1.703	2.052	2.473	2.771	3.421	3.690
28	1.313	1.701	2.048	2.467	2.763	3.408	3.674
29	1.311	1.699	2.045	2.462	2.756	3.396	3.659
30	1.310	1.697	2.042	2.457	2.750	3.385	3.646
40	1.303	1.684	2.021	2.423	2.704	3.307	3.551
60	1.296	1.671	2.000	2.390	2.660	3.232	3.460
120	1.289	1.658	1.980	2.358	2.617	3.160	3.373
∞	1.282	1.645	1.960	2.326	2.576	3.090	3.291

This table is reproduced with the kind permission of the Trustees of Biometrika from E. S. Pearson and H. O. Hartley (eds.), *The Biometrika Tables for Statisticians,* Vol. 1, 3d ed., *Biometrika*, 1966.

Table C
The χ^2-distribution

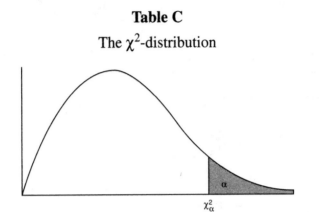

$$\chi^2_\alpha$$

df	$\chi^2_{.995}$	$\chi^2_{.990}$	$\chi^2_{.975}$	$\chi^2_{.950}$	$\chi^2_{.900}$	$\chi^2_{.100}$	$\chi^2_{.050}$	$\chi^2_{.025}$	$\chi^2_{.010}$	$\chi^2_{.005}$
1	.0000	.0002	.0010	.0039	.0158	2.706	3.841	5.024	6.635	7.879
2	.0100	.0201	.0506	.1026	.2107	4.605	5.991	7.378	9.210	10.60
3	.0717	.1148	.2158	.3518	.5844	6.251	7.815	9.348	11.34	12.84
4	.2070	.2971	.4844	.7107	1.064	7.779	9.448	11.14	13.28	14.86
5	.4117	.5543	.8312	1.145	1.610	9.236	11.07	12.83	15.09	16.75
6	.6757	.8721	1.237	1.635	2.204	10.64	12.59	14.45	16.81	18.55
7	.9893	1.239	1.690	2.167	2.833	12.02	14.07	16.01	18.48	20.28
8	1.344	1.647	2.180	2.732	3.490	13.36	15.51	17.53	20.09	21.95
9	1.735	2.088	2.700	3.325	4.168	14.68	16.92	19.02	21.67	23.59
10	2.156	2.558	3.247	3.940	4.865	15.99	18.31	20.48	23.21	25.19
11	2.603	3.053	3.816	4.575	5.578	17.27	19.68	21.92	24.72	26.76
12	3.074	3.571	4.404	5.226	6.304	18.55	21.03	23.34	26.22	28.30
13	3.565	4.107	5.009	5.892	7.042	19.81	22.36	24.74	27.69	29.82
14	4.075	4.660	5.629	6.571	7.790	21.06	23.68	26.12	29.14	31.32
15	4.601	5.229	6.262	7.261	8.547	22.31	25.00	27.49	30.58	32.80
16	5.142	5.812	6.908	7.962	9.312	23.54	26.30	28.85	32.00	34.27
17	5.697	6.408	7.564	8.672	10.09	24.77	27.59	30.19	33.41	35.72
18	6.265	7.015	8.231	9.390	10.86	25.99	28.87	31.53	34.81	37.16
19	6.844	7.633	8.907	10.12	11.65	27.20	30.14	32.85	36.19	38.58
20	7.434	8.260	9.591	10.85	12.44	28.41	31.41	34.17	37.57	40.00
21	8.034	8.897	10.28	11.59	13.24	29.62	32.67	35.48	38.93	41.40
22	8.643	9.542	10.98	12.34	14.04	30.81	33.92	36.78	40.29	42.80
23	9.260	10.20	11.69	13.09	14.85	32.01	35.17	38.08	41.64	44.18
24	9.886	10.86	12.40	13.85	15.66	33.20	36.42	39.36	42.98	45.56
25	10.52	11.52	13.12	14.61	16.47	34.38	37.65	40.65	44.31	46.93
30	13.79	14.95	16.79	18.49	20.60	40.26	43.77	46.98	50.89	53.67
40	20.71	22.16	24.43	26.51	29.05	51.81	55.76	59.34	63.69	66.77
50	27.99	29.71	32.36	34.76	37.69	63.17	67.51	71.42	76.15	79.49
60	35.53	37.48	40.48	43.19	46.46	74.40	79.08	83.30	88.38	91.95
70	43.27	45.44	48.76	51.74	55.33	85.53	90.53	95.02	100.4	104.2
80	51.17	53.54	57.15	60.39	64.28	96.58	101.9	106.6	112.3	116.3
90	59.20	61.75	65.65	69.13	73.29	107.6	113.1	118.1	124.1	128.3
100	67.33	70.66	74.22	77.93	82.86	118.5	124.3	129.6	135.8	140.2

Adapted with permission from *Biometrika Tables for Statisticians*, Vol. 1, 3d ed., Cambridge University Press, 1966, edited by E. S. Pearson and H. O. Hartley.

Table D

Critical levels of r at 5% and 1% levels of significance

df	$r_{.05}$	$r_{.01}$	df	$r_{.05}$	$r_{.01}$
1	.997	1.000	24	.388	.496
2	.950	.990	25	.381	.487
3	.878	.959	26	.374	.478
4	.811	.917	27	.367	.470
5	.754	.874	28	.361	.463
6	.707	.834	29	.355	.456
7	.666	.798	30	.349	.449
8	.632	.765	35	.325	.418
9	.602	.735	40	.304	.393
10	.576	.708	45	.288	.372
11	.553	.684	50	.273	.354
12	.532	.661	60	.250	.325
13	.514	.641	70	.232	.302
14	.497	.623	80	.217	.283
15	.482	.606	90	.205	.267
16	.468	.590	100	.195	.254
17	.456	.575	125	.174	.228
18	.444	.561	150	.159	.208
19	.433	.549	200	.138	.181
20	.423	.537	300	.113	.148
21	.413	.526	400	.098	.128
22	.404	.515	500	.088	.115
23	.396	.505	1000	.062	.081

Reproduced by the courtesy of the author and of the publisher from G. W. Snedecor and W. G. Cochran, *Statistical Methods.* The Iowa State University Press, Ames, Iowa, 1967, Table A11, p. 557.

GLOSSARY

Included here are the definitions of many, but not all, of the terms used in the keys. For formulas, calculations, and examples of these terms, and for terms not listed here, please consult the index for page references.

α-risk The probability of committing a Type I error

Bar graph A visual representation of data in which frequencies of different results are indicated by the heights of bars representing these results.

Binomial probabilities Probabilities resulting from applications in which a two-outcome situation is repeated some number of times, and the probability of each of the two outcomes remains the same for each repetition.

Box and whisker display A visual representation of dispersion which shows the largest value, the smallest value, the median, the median of the top half of the set, and the median of the bottom half of the set.

β-risk The probability of committing a Type II error

Central Limit Theorem Pick n sufficiently large (at least 30), take all samples of size n, and compute the mean of each of these samples. Then the set of these sample means will be approximately *normally* distributed.

Chebyshev's theorem For any set of data, at least $(1 - 1/k^2)$ of the values lie within k standard deviations of the mean.

Chi-square A probability distribution used here in goodness-of-fit tests and for tests of independence.

Confidence interval The range of values that could be taken at a given significance level.

Correlation coefficient A measure of the relationship between two variables.

Critical value A value used as a threshold to decide whether or not to reject the null hypothesis.

Empirical rule In symmetric "bell-shaped" data, about 68% of the values lie within one standard deviation of the mean, about 95% lie within two standard deviations of the mean, and more than 99% lie within three standard deviations of the mean.

Expected value (or **average** or **mean**) For a discrete random variable X, this is the sum of the products obtained by multiplying each value x by the corresponding probability $P(x)$.

Histogram A visual representation of data in which relative frequencies are represented by relative areas.

Mean Result of summing the values and dividing by the number of values.

Median The middle number when a set of numbers is arranged in numerical order. If there are an even number of values, the median is the result of adding the two middle values and dividing by two.

Mode The most frequent value.

Normal distribution A particular bell-shaped, symmetric curve with an infinite base.

Null hypothesis A claim to be tested, often stated in terms of a specific value for a population parameter.

Operating characteristic curve A graphical display of β values.

Outlier An extreme value falling far from most other values.

Percentile ranking Percent of all scores that fall below the value under consideration.

Poisson distribution A probability distribution which can be viewed as the limiting case of the binomial when n is large and p is small.

Population Complete set of items of interest.

Power curve The graph of probabilities that a Type II error is not committed.

Probability A mathematical statement about the likelihood of an event occurring.

P-value The smallest value of α for which the null hypothesis would be rejected.

Random sample When the sample is selected under conditions such that each element of the population has an equal chance to be included.

Random variable Real numbers associated with the potential outcomes of some experiment.

Range The difference between the largest and smallest values of a set.

Regression line The best fitting straight line, that is, the line which minimizes the sum of the squares of the differences between the observed values and the values predicted by the line.

Sample Part of a population used to represent the population.

Scatter diagram A visual display of the relationship between two variables.

Significance level The choice of α-risk in a hypothesis test.

Simple ranking After arranging the elements in some order, noting where in the order a particular value falls.

Skewed A distribution spread thinly on one end.

Standard deviation The square root of the variance.

Stem and leaf display A pictorial display giving the shape of the histogram, as well as indicating the values of the original data.

Student t-distribution A bell-shaped, symmetrical curve which is lower at the mean, higher at the tails, and more spread out than the normal distribution; often used when working with small samples.

Type I error The error of mistakenly rejecting a true null hypothesis.

Type II error The error of mistakenly failing to reject a false null hypothesis.

Variance The average of the squared deviations from the mean.

Z-score The number of standard deviations a value is away from the mean.

INDEX